Studio Drama
Processes and Procedures

Multiple Camera Video Series
By Robert J. Schihl

Television Commercial Processes and Procedures

Talk Show and Entertainment Program Processes and Procedures

Studio Drama Processes and Procedures

TV Newscast Processes and Procedures

Studio Drama
Processes and Procedures

Robert J. Schihl, Ph.D.

Focal Press

Boston London

Focal Press is an imprint of Butterworth–Heinemann.

Recognizing the importance of preserving what has been
written, it is the policy of Butterworth–Heinemann to have the
books it publishes printed on acid-free paper, and we exert our
best efforts to that end.

Library of Congress Cataloging-in-Publication Data

Schihl, Robert J.
    Studio drama processes and procedures / Robert J.
  Schihl.
      p.   cm.—(Multiple camera video series)
    Includes bibliographical references (p.   ) and index.
    ISBN 0-240-80096-6 (pbk.)
    1. Television plays—History and criticism.
  2. Television—Production and direction.   3. Soap operas.
  I. Title.   II. Series: Schihl, Robert J. Multiple camera video
  series.
  PN1992.65.S3 1991
  791.45′0232—dc20   91-16859
                                                            CIP

British Library Cataloguing in Publication Data

Schihl, Robert J.
    Studio drama processes and procedures.—(Multiple
  camera video series)
    I. Title II. Series
    791.45

    ISBN 0-240-80096-6

Butterworth–Heinemann
80 Montvale Avenue
Stoneham, MA 02180

10 9 8 7 6 5 4 3 2 1

Printed in the United States of America

Contents

Preface

Multiple camera video production refers to those varying video projects that create the products of the multiple camera video technology. It is usually the video production associated with a studio setting. Some of the most common multiple camera studio production today includes drama production (mainly soap operas), newscast production, talk show/entertainment production, and television commercial production. Of these, only the daily newscast and occasional television commercial are still being produced at local television production studios. Occasionally, public service talk shows are produced at the local facility.

In network television studio production, besides the daily newscasts, the continuing daily soap opera production still occupies one of the busiest and most productive of the studio projects.

While the basic skills necessary for studio production remain the same across studio production genres, preproduction tasks and production roles vary immensely from one genre to another. The beginning television producer, director, and production crews need to be aware of those studio production elements that are similar across genres as well as those that differ. It is not uncommon to find well-meaning beginners performing poorly in their roles across genres when principles and criteria for one genre were erroneously applied to another. Professionally, even apparently similar studio production roles across television program genres require a different mind set and, consequently, different skills application. Hence, at beginning and intermediate television production levels, awareness of the various genres and their differences encourages varying skill development and consequent employment marketability.

Academic goals for beginning television production skills are usually oriented to some form of studio operation. Therein students have the opportunity to learn all aspects of video production from the technology of the medium to the aesthetics of the final product. In some academic programs, knowledge of these concepts and skills is required before advancing to the single camera video experience. The television studio experience encompasses all phases of video *producing skills* from those of the producer to those of the director and all phases of video *production technology* from that of the camera operator (video camera skills) to that of the technical director (video editing skills). Given success in the learning stages of video production, skills still need expression. The most common expression of video production skills is to engage in some form of television studio production. Beyond a basic television studio production handbook, there is not an adequate text available to design and organize differing studio production genres. This book concentrates on producing and the production of television studio—three camera—drama.

The growth of both broadcast and cable television in the United States and abroad in the 1990s shows ever increasing development and expansion. The areas of low power television (LPTV) production, the continuing increase in cable origination programming, and the introduction of high definition television (HDTV) attest to that development and expansion. The rapid fragmentation of cable and satellite television in general is spurring the proliferation of independent television production facilities and the multiplication of television studio program genres. Producers for these independently produced and syndicated programs seek from media conventions the guidelines for producing and the production of the various genres. Publishers of books on the subject of television are seeing a growing need in these areas and are responding with an increasing number of publications. These publications assist the entrepreneur in the development of video production houses and in the training of producers. What is needed by both the broadcast educator as well as the video production entrepreneur is guidance for producing and production procedures, information on video production organizational flow, and the definition of video producing and production roles.

Following the publication of my earlier book *Single Camera Video: From Concept to Edited Master* (Focal Press, 1989), I found that educators, beginning professionals, and industry reviewers alike applauded the format in which I presented the procedural, organizational, and role definition stages of the single camera video technology. I was asked why such a format was not available for multiple camera video production. Having spent more years both in broadcast education and professional production with multiple camera video production than with single camera video production, I asked myself the same question. I hope this text is the desired response.

Studio Drama Processes and Procedures is intended to meet the needs of those individuals from intermediate television education to the beginning television production facility entrepreneur who needs a clear and comprehensive road map to organizing a television studio drama production. This book presents the conceptual preproduction stages of drama production (e.g., choosing a teleplay; auditioning a cast; and preparing a production budget, studio set design, and lighting and audio designs) as well as all levels of production skill definition and organization until the director calls for a wrap to the studio production. This road map includes the thoroughness of a flowchart for all preproduction, production, and postproduction personnel roles and the specific detail of preproduction and production organizing forms that facilitate and expedite most roles throughout the process of the studio production of television drama.

In *Single Camera Video: From Concept to Edited Master* I referred to the approach I took as "a cookbook recipe approach." I have not only not regretted the comparison but have grown confirmed in the analogy. Similar to single camera video production, multiple camera video production is a process needing the guidance of a good kitchen recipe—an ordered, procedural, and detailed set of guidelines. As with a good recipe, after initial success, a good cook adjusts ingredients, eliminates some, adds others, and makes substitutions according to taste and experience with the recipe. So, too, is it with the approach taken in this text. The production organizing details covered in this text are such that after initial introduction, depending often on available personnel, studio equipment, or production time frame, preproduction and production task roles can be multiplied, combined, or eliminated—just like the ingredients in a favorite recipe.

Specifically, the approach that this text takes to multiple camera video production of television studio drama is one of organization. The most frequent and vociferous compliment to *Single Camera Video* is that it organizes the whole gamut of the preproduction, production, and postproduction of single camera video production. In the same way, this text is designed to organize the multiple camera video production of television studio drama. Books on the other television studio genres (the television newscast, the talk show/entertainment program, and the television studio commercial production) are also available from Focal Press in worktexts similar to this.

The process of television drama production is presented in Chapter 1 in flowchart checklist form for preproduction, production, and postproduction stages of production. For each of the three stages, producing and production personnel roles are presented, with responsibilities and obligations listed in their order of accomplishment. In Chapter 2, the processes of preproduction, production, and postproduction are presented in chronological order of production role performance. In Chapter 3, production organization forms are provided for various preproduction and production tasks. These forms are designed to facilitate the goal or producing and production requirements at certain stages of the process of producing and production. A description of and glossary for each form are presented in Chapter 4 to facilitate use of the forms.

In determining the method of presenting the television studio drama production genre in this text some choices had to be made. Production procedures had to be adapted to a learning process. The first choice was to decide the number of producing personnel and the studio production crew size. The number of producing roles and studio production crew members usually is determined by the production budget and the number of available personnel (i.e., students) in an academic environment. The choice

was made to create above-the-line and below-the-line staff and crew for an average size production with a minimum number of personnel. As in the analogy of a good recipe, combine, subtract, or multiply staff and crew members according to costs and availability. The industry does that, too.

Another choice was to determine the level of studio hardware and technology sophistication to assume in drama production. No two television studio facilities are alike, especially in academic environments. The criterion for level of hardware and technology sophistication assumed was to be aware that no matter how much or how little a studio facility has in terms of hardware and technology, the students leaving their training time in any type of production facility still have to know some point of minimal reference to the "real world" of studio drama production and employment skill expectation standards. A teacher of broadcasting can easily structure this text and its approach to an individual facility. Without an ideal point of reference, it would be very difficult to structure up to a level that was not covered in this text. Students must be made aware of what to expect in a "normal" or "ideal" production facility and from a studio production crew. This is what I have tried to convey. I have taught studio drama production from a minimal facility to a full blown state-of-the-art facility. This text is adaptable to either.

I repeat what I wrote in *Single Camera Video*. The video production industry or television studio drama production is not standardized in the procedural manner in which its video products are created. This text, although it can appear to recommend standardized approaches to achieving video products, is not intended to imply that the industry or studio drama production is standardized nor is it an attempt to standardize the industry or studio drama production. The presentation of ordered production steps is merely academic; it is a way of teaching and learning required stages in video production. The video product is the result of a creative process and should remain that way. This text provides order and organization to television studio drama production for beginning producers and directors and for intermediate television teachers and students of video production. Once someone becomes familiar with the process of any television studio drama production, that person is encouraged to use what works and facilitates a task and to drop or rethink those elements that do not work or no longer facilitate the task.

I am currently a full professor—both a charter and senior television faculty member—at Regent University on the grounds of the Christian Broadcasting Network, home of the Family Channel. I teach in both the School of Radio, Television, and Film, and the School of Journalism, and work with faculty and graduate students of the Institute of Performing Arts.

ACKNOWLEDGMENTS

I am indebted to many people over many years who have contributed unknowingly to this book:

To Joanne, Joel, and Jonathan for putting up with the interminable clacking of computer keys and the whine of a printer;

To my parents, Harold and Lucille, who unwittingly turned me on to television production in 1950 by purchasing our first television set;

To Dr. Marilyn Stahlka Watt, Chair, Department of Communication, Canisius College, Buffalo, New York, for introducing me to television broadcasting and for making me a television producer;

To Edward Herbert and Kurt Eichsteadt, Taft Broadcasting, for giving me the chance to have my own live, local prime-time television program;

To Marion P. Robertson, Chief Executive Officer, the Family Channel, Virginia Beach, Virginia, for giving me the opportunity to work and teach in a state-of-the-art national network television facility;

To Dean David Clark, Provost George Selig, and President Bob Slosser, Regent University, Virginia Beach, Virginia, for granting me the sabbatical to write this book;

To Dr. Elaine Shouse-Waller, Regent University, Institute of Performing Arts, for the memory of our first cooperative venture in studio drama—the first ever studio production in the new Christian Broadcasting Network facility in the summer of 1979;

To Karen Speerstra, Senior Editor, and Philip Sutherland, Acquisitions Editor, Focal Press, Stoneham, Massachusetts, for being the most encouraging editors and friends an author could have;

To Rob Cody, Project Manager; Julie Blim, "700 Club" Producer; and John Loiseides, Photographer, Christian Broadcasting Network, Virginia Beach, Virginia, for their research assistance;

To my thousands of television students from the State University of New York at Buffalo, the State University of New York College at Buffalo, Hampton University, and Regent University, and especially to those among them whose names I see regularly on the closing credits of network and affiliate television programs—for the thrill of seeing their names.

A.M.D.G.
Robert J. Schihl
Virginia Beach, Virginia
1991

Key to the Book

Creating a successful television studio production requires that many tasks at varying stages of production be performed in sequential order. Most tasks build upon one another and are interrelated with the tasks of other production personnel. The television industry has clearly defined roles for each production personnel member. This book provides a blueprint of the duties assigned to each production role. The duties of each role—producer, director, camera operator, etc.—are first presented in a checklist flowchart. Role duties are then cross-referenced within the text on the basis of studio production stages. The tasks necessary for every production role are divided and arranged at each stage of the production process. The flowchart is divided into the three chronological stages of television studio production: preproduction, production, and postproduction.

This book can be used as a text or reference. The reader can gain a comprehensive understanding of the production process by reading the entire book. The reader also can check specific personnel responsibilities or use particular forms.

Chapter 1 presents a flowchart for preproduction, production, and postproduction stages by personnel role and in sequential task order (see figure). Chapter 2 details each production task in the order in which it is to be completed. Chapter 3 provides production organizing forms. Chapter 4 is a key to terms and information for the production forms in Chapter 3.

Following is a listing of the personnel abbreviations used throughout this text.

Personnel	Abbreviation
Producer	P
Director	D
Talent/actor/actress	T
Set designer	SD
Properties master/mistress	PM
Costume master/mistress	CM
Make-up artist	MA
Assistant director	AD
Floor director	F
Technical director	TD
Camera operators	CO
Audio director	A
Lighting director	LD
Continuity person	CP
Production assistant	PA
Microphone boom grip(s)/ operator(s)	MBG/O
Video engineer	VEG
Videotape recorder operator	VTRO

Choose a television studio production stage. Here, the PREPRODUCTION stage was chosen.

PREPRODUCTION

PRODUCER (P)

Each television studio production role is abbreviated (e.g., P) and sequential production tasks numbered (e.g., (1), (2), (3), etc.). To locate the fourth PREPRODUCTION task responsibility of a PRODUCER (P) look down the checklist for (4).

- (1) Selects a teleplay script **11**
- (2) Constructs a budget; authorizes expenses (production budget form) **12**
- (3) Secures/maintains the studio production facility schedule/relationship (facility request form) **12**
- (4) Chooses/meets with the director **12**
- (5) Holds auditions; casts the teleplay with the director (talent audition form) **13**
- (6) Supervises cast interpretive reading (characterization form) **13**

Every television studio production role and task is listed here and is explained in detail in Chapter 2. The page number after the entry directs you to the explanation. Chapter 2 will also aid you in learning the responsibilities of the other crew members.

- (7) Approves the set design(s); authorizes expenses and purchase requisitions **14**
- (8) Approves the properties list; authorizes expenses and purchase requisitions **15**
- (9) Approves the costume designs; authorizes expenses and purchase requisitions **16**
- (10) Approves the make-up designs; authorizes expenses and purchase requisitions **16**
- (11) Approves the audio plot designs and music; authorizes expenses and purchase requisitions **22**

Reading this information will indicate if there is a production organizing form available in Chapter 3 to assist you in a particular task.

- (12) Obtains copyright/royalty clearances **24**
- (13) Secures insurance coverage **24**
- (14) Approves the lighting plot designs; authorizes expenses and purchase requisitions **25**
- (15) Designs the titling/credits (titling/credits copy form) **26**

Each television studio production organizing form is explained in detail in a glossary of terms and information required for the form in Chapter 4.

Annotated flowchart.

Television Drama Production

INTRODUCTION

Survey research companies in the 1980s consistently found in studying the television viewing habits of Americans that their top two viewing choices were news and drama. News and drama (puppets and charades rounded out the two remaining program choices) were the first few limited programming choices in the early years of television.

The Golden Age of Television (from the late 1940s through the 1950s) is distinguished for the extent of live television studio drama produced in those years. "Kraft Television Theater," and "Studio One" of the 1940s, as well as "Playhouse 90," "United States Steel Hour," "DuPont Show of the Month," "Producers Showcase," "The Best of Broadway," "Television Playhouse," "Robert Montgomery Presents," "Ford Star Jubilee," "Hallmark Hall of Fame," and "General Electric Theater" of the 1950s brought live and original full length plays into the homes of Americans each week. The radio shows that turned to television, while remaining essentially comedic, were nonetheless multiple camera play productions of their natures. "The Goldbergs" (1949), "The Life of Riley" (1949), "Amos 'n Andy" (1951), and others were produced as teleplays with multiple studio cameras. Television producers knew that the viewers of the new medium had a voracious appetite even in those days. And the tedious process of single camera film in the production of drama would never have met the demands of the new electronic medium. For television, the use of multiple cameras was the norm in studio production.

The Golden Age was an age, too, when the first television drama producers, directors, and writers knew that this new medium was essentially different from theater and from Hollywood filmmaking. And they acted accordingly. Without videotape recording technology to prerecord any part of a teleplay, those early producers of television drama mounted, staged, and televised full length dramatic endeavors straight through from opening line to closing line. Writers like Paddy Chayefsky, Reginald Rose, Herman Wouk, and Rod Serling knew how this new medium differed from other drama media and wrote for the small screen. Teleplays like "Marty" (1953), "Twelve Angry Men" (1954), "The Caine Mutiny Court Martial" (1955), and "Requiem for a Heavyweight" (1956) became benchmarks for drama and for television. Original teleplays were the norm for the live and Golden Age of television drama.

Today, some of the only remaining television studio drama—and multiple camera production—are the continuing daily drama series known as soap operas. Each of the national networks produces daily soap operas for an as-yet-unsatiated television audience.

FLOWCHART AND CHECKLIST

The following pages present in flowchart and checklist fashion the procedures of the preproduction, production, and postproduction stages of the television studio production of a teleplay. In this section, the procedures are ordered by the definition of personnel role responsibility at the three stages. Before each entry in the procedure, a checklist box permits notation of the completion of each step as preproduction, production, and postproduction processes proceed.

FLOWCHART AND CHECKLIST FOR THE STUDIO DRAMA PRODUCTION

PREPRODUCTION	PRODUCTION	POSTPRODUCTION

PRODUCER (P)

PREPRODUCTION	PRODUCTION	POSTPRODUCTION
(1) Selects a teleplay script **11**	(1) Supervises crew/cast calls; meets with the director and crew/cast **32**	(1) Arranges for postproduction schedule and facilities **47**
(2) Constructs a budget; authorizes expenses (production budget form) **12**	(2) Handles studio/set arrangements **32**	(2) Supervises postproduction preparation **47**
(3) Secures/maintains the studio production facility schedule/relationship (facility request form) **12**	(3) Handles crew and cast details **32**	(3) Observes editing with the director; assists decision making regarding edits **48**
(4) Chooses/meets with the director **12**	(4) Confers with the director on acceptable take(s) **44**	(4) Reviews the rough edit with the director **49**
(5) Holds auditions; casts the teleplay with the director (talent audition form) **13**	(5) Secures talent release signatures (talent release form) **45**	
(6) Supervises cast interpretive reading (characterization form) **13**		
(7) Approves the set design(s); authorizes expenses and purchase requisitions **14**		
(8) Approves the properties list; authorizes expenses and purchase requisitions **15**		
(9) Approves the costume designs; authorizes expenses and purchase requisitions **16**		
(10) Approves the make-up designs; authorizes expenses and purchase requisitions **16**		
(11) Approves the audio plot designs and music; authorizes expenses and purchase requisitions **22**		
(12) Obtains copyright/royalty clearances **24**		
(13) Secures insurance coverage **24**		
(14) Approves the lighting plot designs; authorizes expenses and purchase requisitions **25**		
(15) Designs the titling/credits (titling/credits copy form) **26**		

DIRECTOR (D)

PREPRODUCTION	PRODUCTION	POSTPRODUCTION
(1) Meets with the producer **13**	(1) Holds scheduled production meeting with the crew/cast; sets first unit blocking time **32**	If the director edits:
(2) Attends audition; casts the teleplay with the producer (talent audition form) **13**	(2) Supervises final studio and set arrangements **32**	(1) Supervises postproduction preparation (postproduction cue sheet form) **47**
(3) Schedules the cast reading/interpretation session (characterization form) **13**	(3) Supervises equipment set-up/placement **32**	(2) Reorders/edits master script **47**
(4) Does a script breakdown (script breakdown form) **13**	(4) Begins blocking; blocks actor(s)/actress(es) without lines **36**	(3) Makes a rough edit **48**
(5) Chooses/meets with the set designer; supervises the set design(s); assigns scale models of the set(s) **14**	(5) Rehearses actor(s)/actress(es) without lines; watches actor(s)/actress(es) walk through blocking **36**	(4) Reviews the rough edit **48**
(6) Approves the set design(s) **14**	(6) Rehearses actor(s)/actress(es) blocking with lines **36**	(5) Makes the final edit **49**
(7) Chooses/meets with the properties master/mistress; supervises the properties list **14**	(7) Blocks cameras as cast hits spike marks **37**	(6) Labels the edited master; removes the record button from the videocassette **49**
(8) Approves the properties list **15**	(8) Rehearses cameras and actor(s)/actress(es) **38**	
(9) Chooses/meets with the costume master/mistress; supervises costume selection **15**	(9) Makes changes; rehearses in the studio **38**	If the director does not edit:
(10) Approves the costume designs **16**	(10) Retires to the control room; rehearses from the control room **41**	**TECHNICAL DIRECTOR/EDITOR (TD/E)**
(11) Chooses/meets with the make-up artist; supervises make-up design **16**	(11) Calls a break for costuming/make-up **41**	(1) Edits from the master script and videotape log forms under the supervision of the producer and director **48**
(12) Approves the make-up designs **16**	(12) Calls for a take or another rehearsal after costuming/make-up **42**	(2) Adds music/effects; mixes channels **49**
(13) Creates the master script; designs the blocking plots; uses scale models of set(s); storyboards the master script; plans stock shots and B-roll needs (blocking plot form) **16**	(13) Finishes the take; stops tape **42**	(3) Labels the edited master; removes the record button from the videocassette **49**
(14) Creates the studio production schedule (production schedule form) **19**	(14) Confers with the producer on acceptability of the take; decides/takes pickup shots for editing **44**	
(15) Chooses/meets with the camera operators **19**	(15) Announces intentions: take/wrap/next unit/strike **45**	
(16) Chooses/meets with the audio director; assigns the audio plot **20**		
(17) Approves the audio plot(s) **22**		

☐ (18) Chooses/meets with the lighting director; assigns the lighting plot **25**
☐ (19) Approves the lighting plot(s) **25**
☐ (20) Chooses/meets with the technical director; assigns the shot list breakdown **26**
☐ (21) Chooses/meets with the assistant director and production assistant **26**
☐ (22) Chooses/meets with the floor director and continuity person **26**
☐ (23) Chooses/meets with the videotape recorder operator **29**
☐ (24) Reviews completed set(s) with the set designer **29**
☐ (25) Checks the lighting design(s) and preliminary set lighting **29**
☐ (26) Reviews properties/demonstrations with the properties master/mistress **29**
☐ (27) Reviews the completed costumes with the costume master/mistress and cast **30**
☐ (28) Reviews the make-up tests with the make-up artist **30**

TALENT/ACTOR/ACTRESS (T)

☐ (1) Attends casting audition (talent audition form) **13**
☐ (2) Reads/interprets script with the cast/crew (characterization form) **13**
☐ (3) Learns lines; develops character **13**
☐ (4) Attends fittings; checks costumes; tests make-up designs **16**
☐ (5) Parades costumes before the director for approval **29**
☐ (6) Tries make-up; tests make-up under studio lights and on-camera **30**

☐ (1) Meets with the producer/director for cast call and at the production meeting **33**
☐ (2) Meets the director's schedule demands for lighting stand-in needs **34**
☐ (3) Follows the director's directions for blocking/rehearsal/action/cut/freeze/take **35**
☐ (4) Meets with the make-up artist for make-up preparation; begins make-up application **41**
☐ (5) Meets with the costume master/mistress for costume distribution **41**
☐ (6) Begins costuming **41**
☐ (7) Being available or out of the way during rehearsal(s)/take(s) **42**
☐ (8) Checks with the continuity person between script unit(s)/take(s) **44**
☐ (9) Makes frequent checks with the costume master/mistress and make-up artist to review/repair/touch-up **44**
☐ (10) Removes costumes/make-up; cleans up with strike call **46**

SET DESIGNER (SD)

☐ (1) Meets with the director **14**
☐ (2) Studies the teleplay script **14**
☐ (3) Designs the studio set(s) (set design form) **14**
☐ (4) Submits preliminary design(s) to the producer/director for approval **14**
☐ (5) Meets with the properties master/mistress to review the set design(s) **14**
☐ (6) Constructs scale models of studio set(s) **16**
☐ (7) Begins actual set(s) construction in the studio **18**
☐ (8) Completes set(s) construction; previews set(s) with the director **29**

☐ (1) Oversees the decoration for each set **32**

PROPERTIES MASTER/MISTRESS (PM)

☐ (1) Meets with the director **14**
☐ (2) Meets with the set designer; reviews the set design(s) **15**
☐ (3) Studies the teleplay script **15**
☐ (4) Creates the properties list (properties breakdown form) **15**
☐ (5) Submits the properties list to the producer/director for approval **15**

☐ (1) Meets with the producer/director for cast call and at the production meeting **33**
☐ (2) Completes set(s) decoration with set/hand/action properties **33**
☐ (3) Stands by to handle action properties before and after rehearsal(s)/take(s) **36**
☐ (4) Replaces/replenishes properties consumed/used during rehearsal(s)/take(s) **44**

PREPRODUCTION	PRODUCTION	POSTPRODUCTION
☐ (6) Begins acquisition of the required properties **15** ☐ (7) Reviews/demonstrates collected properties with the director **29**	☐ (5) Replaces/alters properties to the changing needs of the teleplay **44** ☐ (6) Collects/cleans properties after take(s)/strike **46**	

COSTUME MASTER/MISTRESS (CM)

PREPRODUCTION	PRODUCTION	POSTPRODUCTION
☐ (1) Meets with the director **15** ☐ (2) Studies the teleplay script **16** ☐ (3) Designs the costume(s) (costume design form) **16** ☐ (4) Submits the costume design(s) to the producer/director for approval **16** ☐ (5) Acquires/creates/builds costumes **16** ☐ (6) Calls fittings with actor(s)/actress(es); supervises on-camera and lighting tests **16** ☐ (7) Parades the final costumes with the cast before the director **29**	☐ (1) Meets with the producer/director for cast call and at the production meeting **33** ☐ (2) Meets with actor(s)/actress(es); distributes costumes **41** ☐ (3) Assists actor(s)/actress(es) getting into costumes **41** ☐ (4) Watches rehearsal(s)/take(s) for use/abuse/needed repair to costumes **44** ☐ (5) Makes adjustments/repairs to costumes **44** ☐ (6) Retrieves/repairs/cleans costumes after rehearsal(s)/take(s)/wrap(s)/strike **46**	

MAKE-UP ARTIST (MA)

PREPRODUCTION	PRODUCTION	POSTPRODUCTION
☐ (1) Meets with the director **16** ☐ (2) Studies the script **16** ☐ (3) Designs make-up for the cast (make-up design form) **16** ☐ (4) Submits make-up designs to the producer/director for approval **16** ☐ (5) Secures make-up supplies **16** ☐ (6) Tests make-up on actor(s)/actress(es) under lights and on-camera **30** ☐ (7) Reviews make-up designs/tests with the director **30**	☐ (1) Meets with the producer/director for cast call and at the production meeting **33** ☐ (2) Meets with actor(s)/actress(es) for last minute instructions/changes in make-up **41** ☐ (3) Proceeds with make-up application **41** ☐ (4) Checks actor(s)/actress(es) before take(s) **41** ☐ (5) Watches take(s) for make-up needs/changes/repair **41** ☐ (6) Observes studio monitors for additional make-up needs/changes/repair **42** ☐ (7) Knows the rehearsal/take schedule for actor(s)/actress(es); assists in readying actor(s)/actress(es) for subsequent rehearsal(s)/take(s) **44** ☐ (8) Assists actor(s)/actress(es) in removing make-up; assists cleaning up with strike call **46**	

ASSISTANT DIRECTOR (AD)

PREPRODUCTION	PRODUCTION	POSTPRODUCTION
☐ (1) Meets with the director **27** ☐ (2) Secures/purchases adhesive colored dots for spiking actor(s)/actress(es) and cameras **29** ☐ (3) Assigns a color to each actor/actress and camera **29**	☐ (1) Meets with the producer/director for crew call and at the production meeting **33** ☐ (2) Prepares/assists the director with the master script, paperwork, etc. **34** ☐ (3) Checks intercom network connections for the director **35** ☐ (4) Calls/readies actor(s)/actress(es) and cameras by script unit; spikes actor(s)/actress(es) and cameras **35** ☐ (5) Follows the master script for/with the director; notes changes **35** ☐ (6) Makes changes/updates the master script for the director **41** ☐ (7) Follows/keeps track of the master script for the director at the directing console during control room rehearsal(s)/take(s) **42** ☐ (8) Double checks the director's calls for rehearsal(s)/take(s)/wraps/next unit/strike **43** ☐ (9) Assists the director; keeps track of the process; keeps the control room log of takes (videotape log form) **43** ☐ (10) Oversees the strike in the control room; secures the master script and videotape log **45**	☐ (1) Turns the videotape log forms over to the producer **48** ☐ (2) Assists the director during postproduction **48**

FLOOR DIRECTOR (F)

- [ ] (1) Meets with the director **29**
- [ ] (2) Studies the scale model of studio set(s) **29**

- [ ] (1) Meets with the producer/director for crew call and at the production meeting **33**
- [ ] (2) Assumes responsibility for the studio/set/cast/crew during studio use **34**
- [ ] (3) Checks intercom network connections **34**
- [ ] (4) Knows what the director needs/requires during rehearsal(s)/take(s) **34**
- [ ] (5) Effects what the director requires while the director is in the control room **43**
- [ ] (6) Informs self/cast/crew of the director's intentions: retake/wrap/next unit/strike **44**
- [ ] (7) Removes old spike marks before new script unit blocking **44**
- [ ] (8) Oversees the director's call(s): take/retake/wrap/next unit/strike **44**
- [ ] (9) Oversees complete strike of the studio **45**

TECHNICAL DIRECTOR (TD)

- [ ] (1) Meets with the director **26**
- [ ] (2) Becomes familiar with the director's master script **26**
- [ ] (3) Prepares the shot list breakdown from the master script for each camera (camera shot list form) **26**

- [ ] (1) Meets with the producer/director for crew call and at the production meeting **33**
- [ ] (2) Checks intercom network connections **34**
- [ ] (3) Follows the director with a copy of the master script to call the camera shots over the intercom during rehearsal(s)/take(s) from the control room **35**
- [ ] (4) Makes changes/updates the technical director's master script **36**
- [ ] (5) Double checks changes to the master script on each camera's shot list **38**
- [ ] (6) Retires to the control room; responds to the director's calls from the studio over the intercom by cutting shots by consecutive numbering **39**
- [ ] (7) Updates the master script with all changes made during rehearsal critiques by the director **39**
- [ ] (8) Prepares to call all shots by number to camera operators over the intercom during rehearsal(s) when the director retires to the control room **41**
- [ ] (9) Readies all shots in advance by number to camera operators over the intercom during rehearsal(s)/take(s) from the control room **41**
- [ ] (10) Cuts from shot to shot as the director calls for each take during rehearsal(s)/take(s) **41**
- [ ] (11) Assists the director in deciding a take **44**
- [ ] (12) Responds to the director's call: retake/wrap/next unit/strike **44**
- [ ] (13) Shuts down the switcher with wrap/strike call; secures the technical director's master script **45**

CAMERA OPERATORS (CO)

- [ ] (1) Meet with the director **19**
- [ ] (2) Work with the director's master script **19**
- [ ] (3) Study scale model(s) of the studio set(s) **19**

- [ ] (1) Meet with the producer/director for crew call and at the production meeting **33**
- [ ] (2) Set-up/prepare assigned camera/placement **33**
- [ ] (3) Review/attach shot list to the camera **33**
- [ ] (4) Check intercom connections/announce readiness **35**
- [ ] (5) Observe actor(s)/actress(es) blocking/rehearsal by director **36**
- [ ] (6) Are blocked with the actor(s)/actress(es); cameras are spiked **37**
- [ ] (7) Note details of placement/composition/framing; update shot lists **37**
- [ ] (8) Rehearse camera with actor(s)/actress(es) **38**

PREPRODUCTION | PRODUCTION | POSTPRODUCTION

PRODUCTION (continued)

(9) Update camera and actor(s)/actress(es) changes on shot lists **38**
(10) Rehearse as called by the director **39**
(11) Watch for microphone/boom/shadows in shots during rehearsal(s)/take(s) **40**
(12) Shoot camera for take(s) **42**
(13) Prepare to block next unit; strike camera/cables at the director's wrap/strike call **45**

POSTPRODUCTION

(1) Provides music and sound effects tracks (effects/music cue sheet form) **47**
(2) Assists the producer/director with audio problems or additions to the master edit; reviews the rough edit **49**
(3) Sweetens the final audio track for the edited master **49**

AUDIO DIRECTOR (A)

PREPRODUCTION

(1) Meets with the director **20**
(2) Works with the director's master script and set design(s) **20**
(3) Studies scale model(s) of studio set(s) **20**
(4) Designs audio microphone plots (audio plot form) **20**
(5) Designs sound effects/music plot (sound effects plot form) **20**
(6) Submits the audio plot to the producer/director for approval **20**
(7) Chooses/meets with the microphone boom grip(s)/operator(s) **22**

PRODUCTION

(1) Meets with the producer/director for crew call and at the production meeting **33**
(2) Meets with the microphone boom grip(s)/operator(s); begins audio equipment set-up **34**
(3) Runs the microphone cables; records the microphone input(s) **34**
(4) Patches the inputs into the audio control board in the control room **34**
(5) Checks intercom connections with the director, technical director, and audio control board **34**
(6) Places the microphone boom grip(s)/operator(s) with microphone boom(s) for the script unit to be blocked **37**
(7) Makes microphone audio level checks **40**
(8) Makes studio foldback sound check for the microphone boom grip(s)/operator(s) **40**
(9) Rehearses sound effects into the studio **40**
(10) Observes actor(s)/actress(es) blocking; directs microphone boom placement with microphone boom grip(s)/operator(s) **40**
(11) Monitors microphone placement and sound levels during rehearsal(s) **40**
(12) Makes microphone boom grip(s)/operator(s) changes before the next rehearsal **40**
(13) Mixes in sound effects as required **40**
(14) Monitors dialogue and sound effects during rehearsal(s) **41**
(15) Records dialogue and mixes effects during take(s); monitors for any extraneous sounds during take(s) **42**
(16) Prepares for the next script unit with a wrap call by the director; disassembles the microphone/boom/cable equipment; closes down the control board with a strike call **45**

LIGHTING DIRECTOR (LD)

PREPRODUCTION

(1) Meets with the director **25**
(2) Works with the director's master script, set design(s), and properties list **25**
(3) Studies the scale model(s) of the studio set(s) **25**
(4) Designs the lighting plot (lighting plot form) **25**
(5) Submits the lighting plot(s) to the producer/director for approval **25**
(6) Lights the completed studio set(s) **29**
(7) Reviews the set light pattern(s) with the director **29**

PRODUCTION

(1) Meets with the producer/director for crew call and at the production meeting **33**
(2) Lights the set(s) and performs a lighting check; announces readiness **33**
(3) Watches the set(s) and actor(s)/actress(es) during rehearsals for light/shadows **40**
(4) Adjusts lighting/lights after each rehearsal as needed **40**
(5) Studies lighting effects over the studio floor monitors; makes final changes **41**
(6) Monitors the set(s) and actor(s)/actress(es) lighting over the control room monitors during take(s) **42**
(7) Prepares for the next unit; strikes the floor lights; turns off the set lights at strike call **45**

CONTINUITY PERSON (CP)

- □ (1) Meets with the director **29**
- □ (2) Prepares forms for recording continuity notes during rehearsal(s)/take(s) (continuity notes form) **29**

- □ (1) Meets with the producer/director for crew call and at the production meeting **33**
- □ (2) Prepares forms for recording continuity details during rehearsal(s)/take(s) **34**
- □ (3) Records actor(s)/actress(es)/set(s)/dialogue/costume details from unit-to-unit and take-to-take **43**
- □ (4) Releases actor(s)/actress(es) after take(s)/wrap **43**
- □ (5) Establishes previous take details before succeeding take(s)/unit(s) **44**
- □ (6) Reorders continuity forms with wrap/strike **46**

PRODUCTION ASSISTANT (PA)

- □ (1) Meets with the director **27**
- □ (2) Secures titling/credits list from the producer; enters character generator text (titling/credits copy form) **29**

- □ (1) Meets with the producer/director for crew call and at the production meeting **33**
- □ (2) Checks intercom network connections **34**
- □ (3) Enters/records character generator copy for script units to be produced during each production session **34**
- □ (4) Changes relevant information on each slate screen for succeeding takes **42**
- □ (5) Turns the character generator off with wrap/strike call **45**

MICROPHONE BOOM GRIP(S)/OPERATOR(S) (MBG/O)

- □ (1) Meet with the audio director **22**
- □ (2) Work with the director's master script **22**
- □ (3) Design audio pickup plot(s) (audio pickup plot form) **22**
- □ (4) Prepare studio microphone(s)/microphone holder(s)/extension(s) **24**

- □ (1) Meet with the producer/director for crew call and at the production meeting **33**
- □ (2) Meet with the audio director; begin audio equipment set-up **34**
- □ (3) Assemble microphone boom(s) with microphone(s); run audio cables **34**
- □ (4) Take places for preliminary dialogue/sound coverage **37**
- □ (5) Watch actor(s)/actress(es) blocking for microphone coverage; confer with the audio director **39**
- □ (6) Begin sound coverage during camera blocking with actor(s)/actress(es) lines **39**
- □ (7) Confer with the audio director for improved microphone placement; make changes when grip/operator and/or microphone/boom shadows appear on-camera **40**
- □ (8) Note CU/LS framing for placement of microphone(s) for improved audio perspective **40**
- □ (9) Monitor dialogue with foldback/headsets during rehearsal(s)/take(s) **40**
- □ (10) Rehearse with actor(s)/actress(es) and cameras **40**
- □ (11) Make necessary changes/adaptations **40**
- □ (12) Cover actor(s)/actress(es) during take(s) **42**
- □ (13) Prepare for next script unit with wrap call; disassemble microphone(s)/boom(s)/cables with wrap/strike call **45**

VIDEO ENGINEER (VEG)

- □ (1) Does preventive maintenance on cameras **27**
- □ (2) Routes the external signal to the camera monitors for camera operator use **27**

- □ (1) Readies the studio cameras for shading/videotaping **32**
- □ (2) Checks video levels of cameras with the set lighting **33**
- □ (3) Alerts the camera operator and director when ready **33**
- □ (4) Monitors the video level of the cameras during videotaping **42**
- □ (5) Alerts the director to soft focused cameras during videotaping **42**
- □ (6) Caps the cameras with the director's call for a wrap/strike **45**

PREPRODUCTION	PRODUCTION	POSTPRODUCTION

VIDEOTAPE RECORDER OPERATOR (VTRO)

PREPRODUCTION	PRODUCTION	POSTPRODUCTION
☐ (1) Meets with the director **29**	☐ (1) Meets with the producer/director for crew call and at the production meeting **33**	☐ (1) Stripes source/master videotapes with SMPTE time code **48**
☐ (2) Determines videotape stock needs: source tapes and master tape(s) **29**	☐ (2) Prepares the record videotape deck; selects the labeled videotape stock **34**	☐ (2) Turns source tapes and master videotape stock over to the producer **48**
☐ (3) Stripes the SMPTE time code onto stock videotape; codes videotapes; labels videotape stock **29**	☐ (3) Checks intercom network connections **34**	
☐ (4) Determines B-roll videotape needs from the master script **00**	☐ (4) Readies/rolls/records videotape with the director's call **42**	
	☐ (5) Stops videotaping with the director's call **42**	
	☐ (6) Rewinds the record deck and source tape when the taping session is finished; records tape content on the videocassettes **45**	

Chapter 2

Processes
of Television
Drama
Production

INTRODUCTION

There is no one way to produce and videotape television studio drama productions. The three major networks—ABC, CBS, and NBC—differ among themselves in the way they interpret the process of producing television studio drama. The process chosen to be presented in this text is strongly influenced by the ABC–TV model of television studio drama production with some influence from the CBS–TV model.

THE PREPRODUCTION PROCESS

• Personnel

The number of crew personnel is a function of the size of a production facility, the level of technical hardware sophistication, the quality of trained technical crew, the budget, or—in academia—the number of students in a class. The number of necessary crew members in television production can begin with a crew of a tight, basic eight members (consisting of a producer, a director, a technical director, three camera operators, an audio director, and a videotape recorder operator). A crew this small would necessarily involve some doubling up on jobs (e.g., a producer could double as a microphone grip during production and the videotape recorder operator could double as a production assistant).

The length of a teleplay and the size of the cast are also strong considerations in deciding to produce television studio drama. The rule of thumb for determining the videotape length of a teleplay script is to judge one page of the script in accepted teleplay script form as equal to one minute of videotape. The rule of thumb for determin-

ing production length requirements accounts one week (approximately 40 hours) per half hour of edited teleplay.

Ideally, there should be an adequate number of producing staff, including staff to work with the cast as well as staff to prepare the teleplay for production. In this text, a small but adequate preproduction staff will be defined. A cast and its minimum support staff will be included in accounting for a full preproduction personnel staff for television studio drama production.

If the television studio drama production is a commercial venture, then supervisory personnel would hire subordinate staff. In some commercial television studio drama, an executive producer with a teleplay script and finances may begin the process of studio drama production. In an academic environment, supervising faculty may serve as executive producer and may choose the producing staff and production staff or, as is done in some institutions, a producer may be chosen by the faculty, and the student producer chooses qualified peers to round out the above-the-line and below-the-line staff and crew.

In the following role definitions of preproduction personnel, the preproduction meetings called for between the producer and director and their subordinate staff and crew need not be done individually. However, they are defined as individual meetings to stress the importance of communicating with each other during the early stages of production preparation.

Producer (P) The producer in television studio drama normally begins the process of studio drama preproduction. Choosing a script, creating and keeping the budget, choosing a director, obtaining copyright clearances and insurance, and creating titling are among the responsibilities of the producer during preproduction. The qualities of a producer include having good organizational capabilities, being thorough, and paying attention to detail.

FIGURE 2–1
Production personnel organiza-
tional chart. This chart organizes
all production personnel in their
respective supervisory and subordinate positions. The specific skill
category and facility placement are
also indicated.

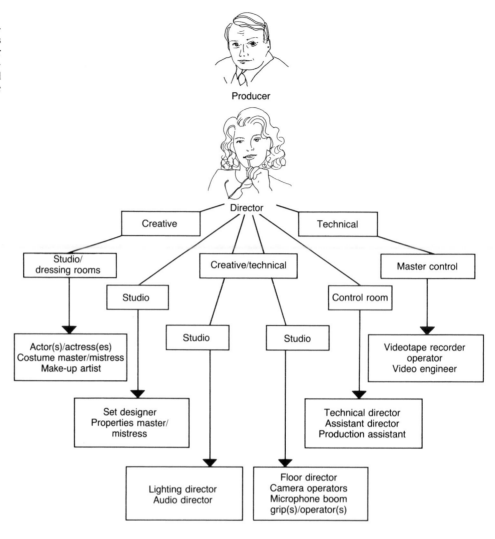

Director (D) The director plays a strong preproduction role as well as the principal production role. All details of studio production preparation fall to the director. Almost all creative and artistic elements of the drama are the responsibility of the director. The qualities of a director include the experience of working with people, directing actors and actresses, and producing television programs.

Camera operators (CO) The camera operators have minimum but important preproduction responsibilities. They have to familiarize themselves with the script, the scale model set(s), and their camera shot lists.

Lighting director (LD) The lighting director's preproduction responsibilities are creative (designing the lighting for the teleplay) and technical (effecting the lighting design with studio light instruments).

Audio director (A) The audio director has preproduction responsibilities ranging from the aesthetic (designing the audio plot and effects for the teleplay) to the technical (physically placing cable runs and microphones for the studio production).

Microphone boom grip(s)/operator(s) (MBG/O) The microphone boom grip(s)/operator(s) spend preproduc-

tion time working out the audio pickup design. By becoming familiar with camera placement and actor(s)/actress(es) blocking, a reasonable idea of microphone and boom placement can be gained.

Assistant director (AD) The assistant director works closely with the director in preproduction, but has minimal preproduction responsibilities.

Floor director (F) The floor director studies the teleplay scale model set(s).

Technical director (TD) The technical director's preproduction responsibilities depend on the director's completion of the master script. The technical director has to create the shot list breakdown for each camera before production can begin.

Videotape recorder operator (VTRO) The videotape recorder operator is not normally considered part of the preproduction crew, but preproduction tasks of a videotape recorder operator are so important that the role is defined here. The videotape recorder operator has the preproduction responsibility for determining the amount of videotape stock necessary for the production and postproduction, and to ready record videotape decks for studio recording. Control room video and audio signals will have to be routed and checked.

Video engineer (VEG) The video engineer is responsible for the proper maintenance of the studio cameras and studio floor monitors. Uncapping cameras, routing the external signal to the cameras, and setting camera lenses to studio set light levels over cameras are all part of the video engineer's preproduction tasks. The more dynamic movement of multiple studio cameras under creative lighting design is an added challenge for the video engineer during teleplay production.

Production assistant (PA) The production assistant's primary preproduction responsibility is to prepare all character generator copy for production—slate, titling, and credits.

Continuity person (CP) The continuity person must prepare a system of recording continuity details in the studio.

Set designer (SD) The set designer is responsible for designing the studio set(s) and overseeing their construction.

Properties master/mistress (PM) The properties master/mistress is responsible for determining and securing all set, hand, and action properties needed for production.

Costume master/mistress (CM) The costume master/mistress is responsible for determining the costume needs

for the teleplay and for obtaining the necessary costumes for the cast.

Make-up artist (MA) The make-up artist is responsible for determining the make-up requirements for each actor and actress in the cast, designing the make-up, and obtaining the necessary make-up supplies.

Talent/Actor/Actress (T) The actor(s) and actress(es) are responsible for working on their characterization and learning the lines for the production. They must also meet make-up design requirements as well as costume measurements and fittings. Make-up and costumes will be reviewed on-camera and under studio lights.

• Preproduction Stages

Making the teleplay script selection (P 1) In most television studio drama production, the producer is the originator of the production (when there is no executive producer). The real point of origination is the selection of a script.

The choice of a script is based on many production details. One consideration is the length of the proposed teleplay. A rule of thumb for determining the length of a teleplay is that one page of script (typed in acceptable script format) is equal to one minute of completed video. If the teleplay will be aired, broadcast time slots have to be considered. The need for commercial break time will

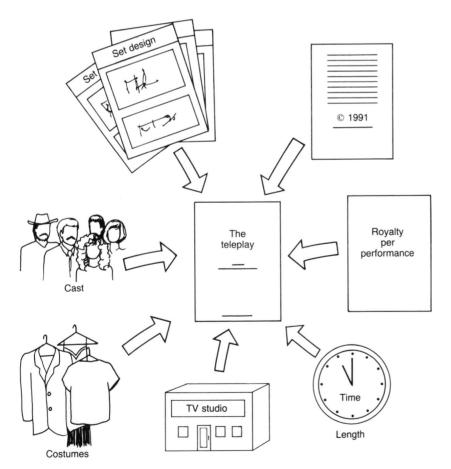

FIGURE 2–2
Teleplay choice considerations. Many elements enter into the choice of a teleplay for studio production.

have to be subtracted from the total time of a videotaped teleplay to accurately determine available telecast length.

Another consideration in the selection of a teleplay in terms of studio space and construction cost is the number of sets a teleplay requires.

Finally, a major consideration in choosing a teleplay is the number of available cast and crew.

Constructing a budget and authorizing expenses (P 2) The producer is responsible for constructing a budget for the production. Depending on the finances available and the extent of the requirements of the script (e.g., number of sets, size of cast, copyright clearance and royalty payments, costuming costs, facility costs, and cast and crew salary requirements), the estimated budget must be as accurate as possible before costs are incurred. There is no doubt that the costs for producing television drama are great and these costs have to be weighed before production begins. This is especially true for a commercial venture. In academia, many costs are subsumed into production expenses for courses. However, many incidental costs are readily incurred and must be accounted for. It is good training in an academic setting for a producer to have to account for all costs for a studio drama production. Students in television drama production must have an understanding of and an appreciation for the real costs of television studio drama outside the academic institution. The producer must approve all preproduction designs and, in so doing, authorizes all expenses to be

incurred by proceeding with the approved designs. (See the production budget form.)

Securing and maintaining the production facility schedule and relationships with the studio supervisors (P 3) The producer begins securing and maintaining a relationship with the studio production facility. Scheduling the teleplay within the studio production facility will be the most important task at this stage. Most studio production facilities do more than one production and need to schedule time as far in advance as possible to accommodate all productions. This is true for both commercial and academic studio facilities. The producer is responsible for maintaining good production facility relationships and communication.

As soon as the teleplay production schedule is created by the director, exact dates and times must be firmed up by the producer with the production facility. Scheduling facility time for set construction and lighting design is critical to ensuring facility availability.

Choosing a studio director and meeting with the director (P 4) Whether in a commercial or an academic venture, the producer must choose a director. The requirements of a director in television drama differ from the same role title in other television genres as well as in other drama media (e.g., proscenium theater). The director in television drama is at once the person directing the teleplay from a television studio perspective as well as

FIGURE 2–3
The producer's preproduction responsibilities. A producer has many important preproduction tasks that must be accomplished early.

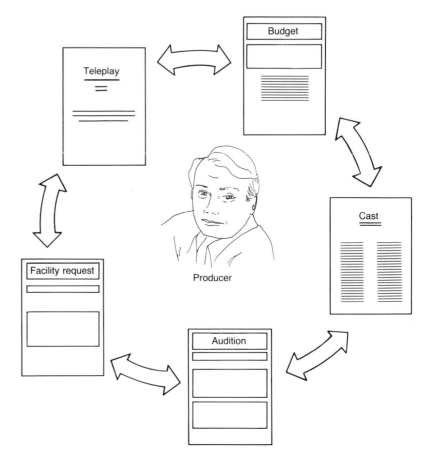

directing the actor(s) and actress(es) from a theater drama perspective. In some productions, a *drama director* may prepare the actor(s) and actress(es). The drama director is not a director in the theater sense of blocking and rehearsing the actor(s) and actress(es). In some ways, a *drama coach* is better suited to the role than a drama director in the theatrical sense.

As soon as a director is hired or chosen, the producer meets with the director and shares the teleplay script, production expectations, and plans.

Meeting with the producer (D 1) The director meets with the producer as soon as possible after being chosen. The director has to be informed of teleplay choice, production expectations, and plans. These plans usually entail production facility use, anticipated production schedule, and deadlines. Other potential crew members are also part of the agenda of this meeting.

Holding auditions for prospective actors and actresses and casting the teleplay with the assistance of the director (P 5) The teleplay gets a running start when the producer schedules auditions for the cast. Auditions are usually posted with enough lead time to attract potential actors and actresses. Auditions should be videotaped so review of potential talent can be made outside the audition itself. The director is a primary consultant and, assuming the director has drama production experience, the casting choices may be the director's. Final casting should be the mutual choice of the producer and director. In a commercial drama venture, costs incurred with union organizations and actors equity have to be weighed. In academia, relations with theater departments and acting majors are also a consideration. In some amateur productions, actors and actresses can be attracted to the cast for the opportunity to have a television studio drama production on their resumé or vita. (See the talent audition form.)

Auditioning for roles in the teleplay (T 1) Prospective talent—actors and actresses—audition for roles in the teleplay. Talent attend an audition expecting to be videotaped. There are many ways to audition for roles in a teleplay. One way involves reading parts of the play for the role(s) best fitting an actor or actress. Other ways are to improvise or characterize. Prospective talent attend an audition prepared with a photograph, a resumé or vita listing previous experience, and information on membership in or affiliation with any unions or equity organization. Talent will be expected to complete an audition information form. (See the talent audition form.)

Casting the teleplay with the producer (D 2) The director assumes an early responsibility to preproduction by sharing the auditioning and casting role with the producer. In some cases, the producer will defer the choice of a cast to the director. Or, depending on the drama experience of the producer, the producer might assume the responsibility of casting. In most cases, it is a shared responsibility and the final cast decisions are mutually agreed upon.

Scheduling a cast reading and interpretation session (D 3) Once the teleplay is cast, the director schedules a reading of the script with the cast. This reading is followed by a character interpretation session during which the director and the cast discuss the production plans and an interpretation of the teleplay and individual characters. (See the characterization form.)

Supervising the cast reading and interpretation (P 6) The producer supervises the cast's scheduled reading and interpretation of the script and the characters. (See the characterization form.)

Reading and discussing the interpretation of the teleplay (T 2) The actor(s) and actress(es) attend a cast reading session with the producer and director. The reading session is a good forum for going over the script once together. Following the reading, a discussion allows the director and cast to talk about the interpretation of the teleplay, the development of the characters, and characterization. If other production staff and crew are available for the reading, they should take the opportunity to attend the reading and interpretation session. (See the characterization form.)

Learning lines and developing characterizations (T 3) The actor(s) and actress(es) spend their remaining preproduction time learning their lines and developing the characterizations required by their roles. The actor(s) and actress(es) will not be involved in theater style rehearsals because blocking will not occur until the production gets to the studio.

Blocking, so common to theater preproduction, is not done this early in television studio drama production because it cannot be accomplished without cameras and camera operators (assuming also the sets and props are in place). Blocking actor(s) and actress(es) in television studio drama production is built into the time and process of production. The only rehearsals that actor(s) and actress(es) should engage in are sit-down reading sessions, common line learning and delivery sessions. Blocking is futile before production begins. Cameras and camera operators are necessary to studio drama blocking because actor/actress blocking is meaningless without the position and lens framing of the cameras. When blocking does occur in the studio, the talent and cameras are blocked together.

Doing a script breakdown (D 4) The director prepares a script breakdown, which is a very important stage in preproduction and upon which much depends. A script breakdown is a listing of all units (or having to decide on the units for videotaping); determining the number of sets, their interior or exterior requirements, the playing locations of those sets, and the cast required for each unit; and, finally, determining the studio videotaping order of shooting the teleplay.

Plays not written as teleplays may have to be reorganized into shooting units to facilitate production. These units are created to make multiple studio camera movement and audio microphone pickup manageable in the studio. The director also needs to be sensitive to the dramatic flow of the script in choosing production units as well as to the anticipated continuity problems in editing the units in postproduction. A scene in traditional theater may have to be broken into multiple units for television studio production. The script breakdown organizes the teleplay into these shooting units and places them in the order in which the units will be produced. (See the script breakdown form.)

Choosing a set designer and meeting with the set designer, supervising the set design(s), and assigning the scale model set(s) (D 5)   After the director chooses a set designer the director meets with the set designer to elaborate on the production requirements and to assist the set designer in beginning set design responsibilities. An important task for the director and set designer is to schedule deadlines for the completion of the set design. The director assigns the set designer a separate deadline to create scale model(s) of the set(s) to be used in blocking the cast and cameras during the director's preproduction work on the master script.

Meeting with the director (SD 1)   The set designer meets with the director as soon as possible after being chosen.

Studying the script (SD 2)   The set designer must read and study the script. The script will suggest a functional set design that will serve the production. The set designer will have to know of and work within any studio facility restrictions (e.g., height of ceiling, lighting battens, and size of the studio).

Beginning the design of the set(s) (SD 3)   The set designer begins as soon as possible to design the required set(s). The set designer works with both the front view and the bird's eye view of each set on some proportional scale. (See the set design form.)

Submitting the set design(s) to the producer and director for approval (SD 4)   The set designer submits the preliminary designs for the teleplay set(s) for approval by the producer and director before proceeding with the scale model mockup miniature(s) or real construction of the set(s).

Approving the set design(s) (D 6)   The director makes any necessary changes or approves the preliminary design(s) before the set designer proceeds with actual construction. The director requires the set design(s) before the master script can be created.

Approving the set designs and authorizing expenses and purchase requisitions (P 7)   The producer reviews the set designs, approves them or makes changes, and authorizes needed expenses and purchase requisitions for set design and construction.

Choosing and meeting with the properties master/mistress and supervising selection of properties (D 7)   The director chooses and meets with the properties master/mistress. The properties master/mistress has to work closely with the set designer. The director shares the aesthetic expectations and plans for the production and sets deadlines for the acquisition of properties.

Meeting with the director (PM 1)   The properties master/mistress meets with the director soon after being chosen by the director. The properties master/mistress needs the director's input in the listing, selection, and acquisition of properties.

Meeting with the properties master/mistress (SD 5)   Once the set designer has approval from the producer and director for the set design(s), the set designer meets with the properties master/mistress before the properties master/mistress begins properties listing and design.

FIGURE 2–4
Scale models of the studio set(s). Many directors find most preproduction tasks easier with the aid of scale models of the studio set(s).

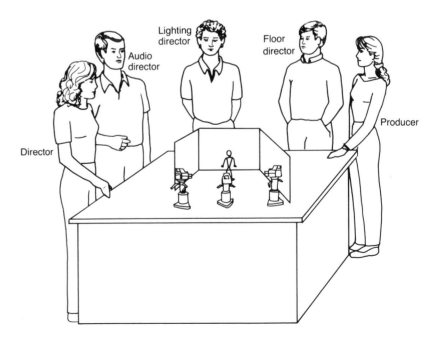

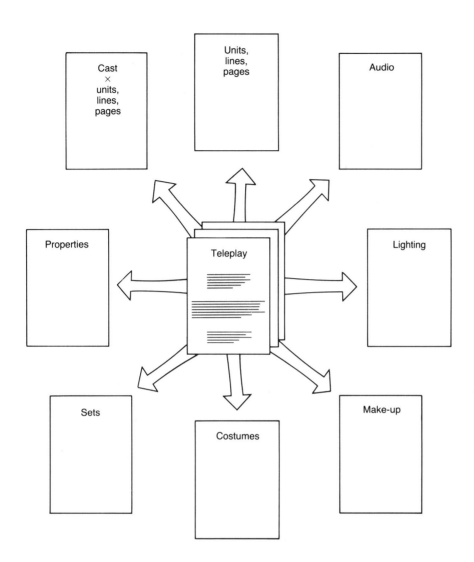

FIGURE 2–5
Teleplay plots and breakdowns. Principal preproduction tasks involve preparing the script breakdown (cast, set(s), and properties) and creating plot designs (audio and lighting).

Meeting with the set designer (PM 2) The properties master/mistress meets with the set designer with the approved set design(s). The set design(s) assist the properties master/mistress in evaluating the properties for the teleplay.

Studying the script (PM 3) With the set design(s) in hand, the properties master/mistress studies the script.

Creating a properties plot or list (PM 4) The properties master/mistress uses the script to create an exhaustive properties plot or list of all set, action, and hand props for the production. (See the properties breakdown form.)

Submitting the properties list to the producer and director for approval (PM 5) The properties master/mistress submits the completed properties list to the producer and director for approval.

Approving the properties list (D 8) The director reviews, makes any changes necessary, and approves the properties list.

Approving the properties list and authorizing expenses and purchase requisitions (P 8) The producer reviews, makes suggestions and changes, and approves the properties list for the production. The producer then authorizes expenses and purchase requisitions for the acquisition of properties.

Beginning the acquisition of properties (PM 6) The properties master/mistress begins to acquire the properties listed on the properties list as approved by the producer and director. Acquisition may require renting, purchasing, or borrowing properties. Borrowing loaned properties may require insurance coverage.

Choosing and meeting with the costume master/mistress and supervising costume selection or design and creation (D 9) The director chooses a costume master/mistress to oversee costuming. The director shares production expectations and planning as well as teleplay interpretation and characterization as an assist in the design or selection of costumes or in the creation and building of costumes.

Meeting with the director (CM 1) The costume master/mistress meets with the director before proceeding with costume design. The director is the resource person with insight into the aesthetic approach and interpretation of the script. An aesthetic approach and script interpretation will translate into appropriate costume choices (i.e., of period or color schemes).

Studying the script (CM 2) The costume designer studies the script as the next step to designing costumes. A script breakdown of costume needs and characters will result from studying the script.

Designing the costumes (CM 3) The costume master/mistress begins to design costumes for individual characters and changes of costumes for characters. Costume design forms that suggest elements of costume design to be considered should be used. (See the costume design form.)

Submitting costume plots to the producer and director for approval (CM 4) The costume master/mistress submits the costume plots to the producer and director for approval.

Approving the costume designs (D 10) The director reviews, makes any changes necessary, and approves the costume designs for the production.

Approving costume designs and authorizing expenses and purchase requisitions (P 9) The producer reviews the costume designs, makes any necessary changes, and approves the costume designs. The producer then authorizes any expenses and purchase requisitions for costume construction or rental.

Choosing and meeting with the make-up artist and supervising make-up design (D 11) The director appoints and meets with the make-up artist. The director shares production requirements for make-up with the make-up artist as well as production expectations, plans, teleplay interpretation, and characterization of roles.

Meeting with the director (MA 1) The make-up artist meets with the director as the first step in the design of make-up for the characters in the teleplay.

Studying the script and characterization (MA 2) The make-up artist studies the script and the characterization of the roles in preparation for beginning make-up design.

Designing make-up (MA 3) The make-up artist uses the make-up design and preparation forms and creates make-up designs and design changes for each character. Part of designing make-up includes determining make-up products to be purchased. (See the make-up design form.)

Submitting the make-up designs to the producer and director for approval (MA 4) The make-up artist submits all make-up designs to the producer and director for final approval.

Approving make-up designs (D 12) The director reviews the make-up designs, makes any additions or changes, and approves the designs.

Approving make-up designs and authorizing needed expenses and purchase requisitions for make-up supplies (P 10) The producer reviews all make-up designs,

makes necessary changes, and approves the designs. The producer then authorizes any necessary expenses and purchase requisitions for make-up supplies.

Securing the required make-up supplies (MA 5) With the approval of the producer and director, the make-up artist begins purchasing or collecting the make-up supplies and materials for the production.

Constructing the scale model set(s) (SD 6) The set designer constructs the scale model(s) of the set(s) before beginning actual set construction. A normal scale used in model set construction is allowing one inch to equal one foot of real space. This standard allows the purchase of one inch to one foot scale human models, furniture, and hand and set properties, which are available from doll-house retailers.

Acquiring and building approved costumes for the production (CM 5) With costume design approval from the producer and director, the costume master/mistress begins to acquire by rental; borrowing; or building the costumes by purchasing needed materials, renting sewing equipment if necessary, and putting together a crew for working on the costumes.

Calling fitting appointments with the cast whenever necessary and testing the costumes under lights on-camera (CM 6) The costume master/mistress calls cast fittings for costumes occasionally during costume construction. At some point during construction, costumes will have to be judged under studio lights and over the camera monitors for their true representation in video.

Attending scheduled fittings and dress rehearsals (T 4) The cast is available for all scheduled fittings for costumes. The cast also parades the costumes in dress rehearsals under studio lights and over the camera monitors before production begins.

Creating the master script, designing the production storyboard, and planning stock shots and B-roll videotape needs (D 13) With the set design(s) and the scale model(s) of the set(s) in hand, the director prepares the master script—a major preproduction task and a pivotal preproduction document. The master script is the script from which the teleplay will be directed and from which most other preproduction stages will be taken (e.g., shot list and lighting design).

The master script contains a page-by-page combination of two elements. The first element for every page is a left-hand column of the script. The second is a three-column right-hand column. The right-hand column is a series of aspect ratio frames, three across, which correspond to each of the three studio cameras. These camera columns are placed to the right of the column containing the corresponding script dialogue. The director may have to prepare master script pages from the original script by photocopying the script to a left-hand column for ease and convenience in designing the master script. In the right-hand three columns the director sketches in the storyboard frames for each camera change to be effected

```
                                                                    8.

34    INT. BOB'S LIVING ROOM - AT DOOR - DAY

      As police officer enters. Joanne closes the door.

                              JOANNE
                              (over)
                    He's really been upset over
                    everything, it's a nightmare ...

                              OFFICER
                    I can understand.

      Joanne goes to bedroom door:

                              JOANNE
                    Bob, Sgt. Hayes' here.

35    INT. BEDROOM - ANGLE ON BOB - DAY

      Lying on the bed, dressed.  He opens his eyes.

                              BOB

                    What's he want?

                              JOANNE
                    Says he wants to see you.
                    I told him you were resting ...

      Bob nods, pulls himself together, gets up.  We sense his
      fear.  He exits.

36    INT. BOB'S LIVING ROOM

      as Bob enters, shakes hands with the officer.

                              OFFICER
                    Sorry to have to bother you,
                    Bob, but since you're leaving
                    tomorrow --

                              BOB
                    It's okay. What's up?

                              OFFICER
                    We've just learned that the slug
                    that killed Mark came from the
                    same .38 Colt automatic that killed
                    another truck driver last month
                    over in Colorado.
                              (beat)
                                                    (CONTINUED)
```

FIGURE 2–6
Teleplay script page. This is a sample teleplay script page as defined in the medium. Note the script page number in the upper right-hand corner. Note also the script unit numbers along the left-hand margin (i.e., 34, 35, 36).

during production. Each storyboard sketch will be drawn next to the script dialogue to which it refers.

Before actually deciding on and sketching the storyboard frames, the director designs the blocking plots containing bird's eye views of all of the set design(s). The director sketches the proposed blocking of all set props and the actor(s) and actress(es) and their movement on the bird's eye view of the set(s). The three studio cameras will also have to be placed in position on the set(s). The director then blocks the actor(s), actress(es), and cameras by using circles with a character's initial within the circle. The circles are repeated at each character's rest position. Arrows connect a character's circles to indicate the character's movement around the set. When a bird's eye view drawing becomes too complicated with circles and arrows, blocking should be continued on another blocking plot form. It is recommended that the director photocopy multiple copies of the set design(s) at the start of the blocking stage.

In preparing a master script, the director places all script and storyboard columns on the right side of a loose-leaf notebook and a page of the blocking plot on the left side facing the script with which it corresponds. Blocking plots are photocopied for successive pages of script using the same blocking and are placed opposite the corresponding script.

With the blocking plots completed, the director returns to the script pages. Each planned camera change is now storyboarded. A simple sketch, often consisting of stick figures with circles for heads, is created in successive storyboard frames corresponding to the script dialogue or stage directions. The director then decides for each storyboard frame (1) a camera (by number: 1, 2, 3) and placement (left to right), (2) the framing of the shot (noted XCU, CU, MS, LS, XLS), (3) any camera movement during the shot (e.g., zooming, arc, trucking, dollying, or pedestaling), and (4) the content of the framing (e.g., M for the character Mary). Each storyboard frame is then numbered (across the three cameras) consecutively in the order in which it will be taken by the camera during videotaping. Some directors skip every tenth number, leaving a number to be able to insert additional storyboard frames later, perhaps during studio blocking. Finally, each storyboard frame is related by a horizontal line drawn to the point in the script where the camera shot is to be changed. A slash is placed in the script at the exact point of the proposed camera change.

Directors are reminded that in dialogue cutting in

FIGURE 2–7
The development of the direc-tor's master script. The master script is *the* production document. It begins with the original script pages, which are photocopied to a left-hand column. Storyboard frames are added, three across, to the right-hand column. To these pages is added the blocking plot to form the master script.

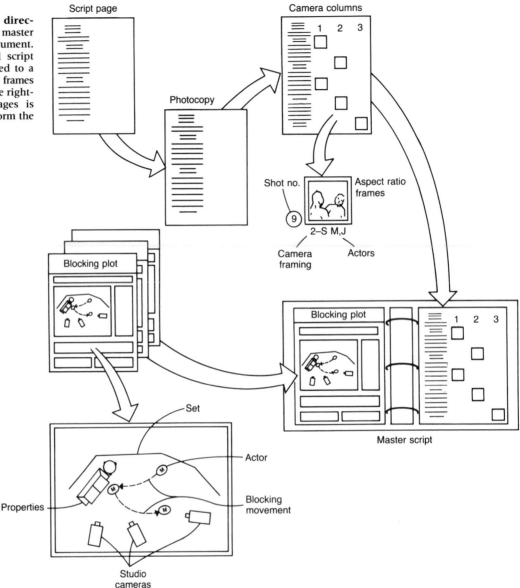

television production, camera changes (e.g., switcher cuts) are taken *before* the last word of the line a character is speaking at an anticipated camera change. This is done for aesthetic reasons, in order to have the character responding in dialogue on-camera as the characters continue the dialogue. Vertical line arrows can link successive storyboard frames to long passages of script for which a particular shot is to be held.

A good aid in designing the blocking plot for the master script is the scale model set(s). The scale model set(s) should be proportional with the set props and three miniature cameras. Camera placement with models of actor(s) and actress(es) will make storyboard design decisions faster and more accurate.

Before the master script is completed, the director has to determine what stock shots will be needed for the completed teleplay. Stock shots are usually exterior scenes used to set a place or atmosphere for the studio interior set(s). Stock shots may be cityscapes, city streets, or house

exteriors. Stock shots may be already available in a videotape library or may have to be videotaped, usually by a single camera.

Finally, the director has to know from the storyboard frames of the master script what units will have to be videotaped on a separate (B-roll) videotape source because a visual transition other than a cut is planned. If B-roll video sources are not preplanned, a second generation of one of the video sources has to be made, which reduces the quality of the video needed for the transition. Preplanning a B-roll videotape eliminates that video quality loss.

Building the approved set design(s) (SD 7) The set designer begins the construction of the set(s) with the completion and approval of the scale model set(s). Set construction takes place either in a special workshop or in the studio. Eventually, studio facility time will have to be scheduled through the producer for in-studio set construction.

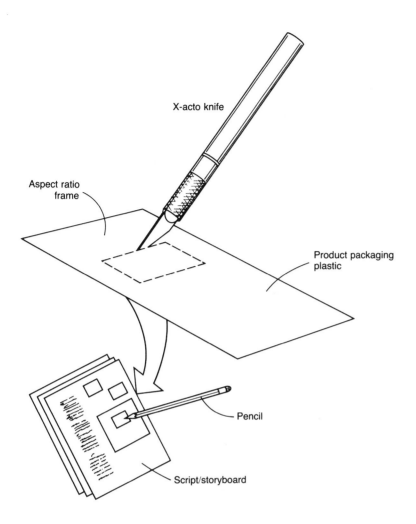

FIGURE 2–8
Aspect ratio frame template. An aspect ratio frame template is easily made by cutting a 3 × 4 unit rectangle from plastic product packaging with an x-acto knife.

Creating a production schedule and studio calendar (D 14) As soon as the director completes the master script, information is available for planning a production schedule and calendar of studio use. Production schedules are usually designed around the set(s) that are up in the studio and the availability of actor(s) and actress(es). It is cost effective to videotape at one time all units of the script in any one set than it is to put up and take down the set repeatedly to follow some other order of production. It is also economical to group production units for videotaping around the same number of actor(s) and actress(es) available at the same time. It is hard to justify having an actor or actress present for a full production period for only one line if it can be avoided. Some other order of production could schedule that actor or actress with other units in which they appear in order to make more efficient use of their time.

The production schedule translates into a calendar of production dates, hours, crew and cast calls, script units, sets, and approximate strike times. (See the production schedule form.)

Choosing and meeting with the camera operators (D 15) The director chooses and meets with the three camera operators. Choosing good camera operators is essential to the success of the production. Each camera will eventually be blocked, and the camera operator learns blocking just as an actor or actress learns blocking. In addition to blocking, every camera shot assigned to a camera, framing, shot composition, and any camera movement must become second nature to the camera operator. And every camera operator has to be a videographer—a photographer with a video camera. The director makes a photocopy of the master script available to the camera operators and helps them prepare for production.

Meeting with the director (CO 1) The camera operators meet with the director to gain some preproduction insight into the aesthetic nature of the teleplay as envisioned by the director. The camera operators review the director's expectations during production. Any difficult shots or camera movements expected during production can be introduced and reviewed.

Working with a copy of the master script (CO 2) The camera operators prepare for production by reviewing a copy of the director's master script. The master script gives the camera operators a better familiarity with the director's expectations of them and the demands of the script and production.

Studying the scale model(s) of the studio set(s) (CO 3) Camera operators will find it beneficial to study the scale model(s) of the studio set(s) for the teleplay during preproduction. Studying the set(s) in the scale

FIGURE 2–9
Photocopied teleplay script page.
This is a sample of a script page photocopied to the left-hand column of a page.

```
34    INT. BOB'S LIVING ROOM - AT DOOR - DAY                              8.

      As police officer enters.  Joanne closes the door.

                              JOANNE
                              (over)
                   He's really been upset over
                   everything, it's a nightmare …

                              OFFICER
                   I can understand.

      Joanne goes to bedroom door:

                              JOANNE
                   Bob, Sgt. Hayes' here.

35    INT. BEDROOM - ANGLE ON BOB - DAY

      Lying on the bed, dressed.  He opens his eyes.

                               BOB

                   What's he want?

                              JOANNE
                   Says he wants to see you.
                   I told him you were resting …

      Bob nods, pulls himself together, gets up.  We sense his
      fear.  He exits.

36    INT. BOB'S LIVING ROOM

      as Bob enters, shakes hands with the officer.

                              OFFICER
                   Sorry to have to bother you,
                   Bob, but since you're leaving
                   tomorrow --

                               BOB
                   It's okay. What's up?

                              OFFICER
                   We've just learned that the slug
                   that killed Mark came from the
                   same .38 Colt automatic that killed
                   another truck driver last month
                   over in Colorado.
                              (beat)                           (CONTINUED)
```

model will begin to orient the camera operators to the studio environment they will find when production begins.

Choosing and meeting with the audio director (D 16) The director chooses and meets with the audio director for the production. The meeting is a chance for the director to convey sound aesthetic expectations. The director also shares expectations and plans for the production. The director gives the audio director a photocopy of the master script to assist in designing the audio plot.

Meeting with the director (A 1) The audio director meets with the director as a first step in preproduction for sound coverage for the production. The audio director comes away from the meeting with a sense of the director's sound aesthetic expectations for the production as well as some idea of the expected music and sound effects. A copy of the master script with the blocking plots will assist the audio director in designing the audio plot(s).

Working with the master script and set design(s) (A 2) The audio director works with the copy of the master script and studies the set design(s) as a prelude to designing the audio plot(s) for the production.

Studying the scale models of the set(s) (A 3) The audio director can learn a lot by studying the scale models of the studio set(s).

Designing audio plot(s) (A 4) Using the audio plot form, the audio director designs the audio plot(s) for the production. From the audio plot(s), the audio director can determine microphone needs, cable needs, and the number of microphone boom grips needed for production. The number of grips and microphone holder equipment can be determined from the distance between actor(s) and actress(es) during dialogue and blocking. (See the audio plot form.)

Designing sound effects and music requirements (A 5) The audio director designs the sound effects required during production and determines the music necessary for the sound track. Part of determining the sound effects and music is alerting the producer to the copyright clearances required to legally use the resources.

Submitting the audio plot(s) to the producer and director for approval (A 6) The audio director submits the audio plot(s) with their hardware, sound effects,

34   INT. BOB'S LIVING ROOM – AT DOOR – DAY   8.

As police officer enters. Joanne closes the door.

JOANNE
(over)
He's really been upset over
everything, it's a nightmare …

OFFICER
I can understand.

Joanne goes to bedroom door:

JOANNE
Bob, Sgt. Hayes' here.

35   INT. BEDROOM – ANGLE ON BOB – DAY

Lying on the bed, dressed. He opens his eyes.

BOB

What's he want?

JOANNE
Says he wants to see you.
I told him you were resting …

Bob nods, pulls himself together, gets up. We sense his
fear. He exits.

36   INT. BOB'S LIVING ROOM

as Bob enters, shakes hands with the officer.

OFFICER
Sorry to have to bother you,
Bob, but since you're leaving
tomorrow --

BOB
It's okay. What's up?

OFFICER
We've just learned that the slug
that killed Mark came from the
same .38 Colt automatic that killed
another truck driver last month
over in Colorado.
(beat)   (CONTINUED)

FIGURE 2–10
Storyboard frames added to the script. Three storyboard aspect ratio frames are added to the right-hand column, three across, with one storyboard representing one of the three studio cameras.

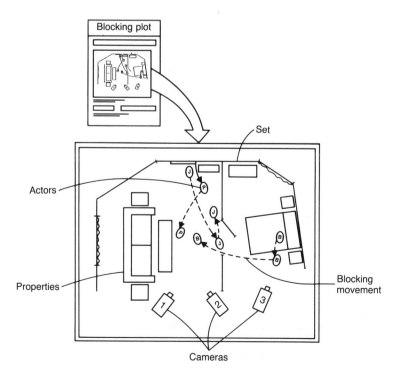

FIGURE 2–11
The blocking plot. The director creates the blocking plot by using the set designer's bird's eye view of the set and adding the cast and their movement, set properties, and the studio cameras and their movement.

FIGURE 2–12
Master script storyboard sample page. Each storyboard frame is placed adjacent to the script dialogue that the frame images.

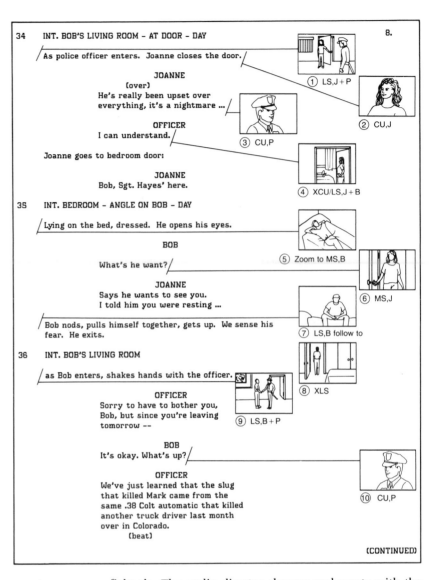

music, and copyright clearance requirements to the producer and director.

Approving the audio plot(s) (D 17) The director reviews the audio plot(s), makes changes or suggestions, and approves the plot(s).

Approving the audio plot(s) and authorizing expenses and purchase requisitions (P 11) The producer reviews the audio plot(s), makes necessary changes, and approves the plot(s). The producer authorizes any necessary expenses and purchase requisitions. The producer notes the music requirements for the teleplay and the necessity of obtaining copyrights and performance clearances.

Choosing and meeting with the necessary number of microphone grip(s)/operator(s) (A 7) As soon as the audio plot(s) are approved, the audio director knows how many studio microphone grip(s)/operator(s) will be needed for the production. If there are platform booms, at least two grips will be needed per platform. One grip will operate the boom with the microphone and the other will have to move the platform. If sound coverage is to be done with fishpole boom(s), one grip will be needed per fishpole. The audio director chooses and meets with the microphone grip(s)/operator(s).

Meeting with the audio director (MBG/O 1) The microphone boom grip(s)/operator(s) meet with the audio director to review audio design plot(s) and plan sound coverage of the studio set(s) and cast. Assignments should be made for particular units; troublesome shooting units are discussed and solutions suggested.

Working with the master script and audio plot design(s) (MBG/O 2) The microphone boom grip(s)/operator(s) study the master script and audio plot design(s). The more familiar the grip(s)/operator(s) are with the master script and audio plot(s), the better prepared they will be for production. Some professionals think that studio microphone grip(s)/operator(s) do even better at performing sound coverage during production if they memorize the script for the cast they cover.

Designing audio pickup plot(s) for production (MBG/O 3) The microphone boom grip(s)/operator(s) design sound pickup requirements for the production. The master script and audio plot(s) should suggest posi-

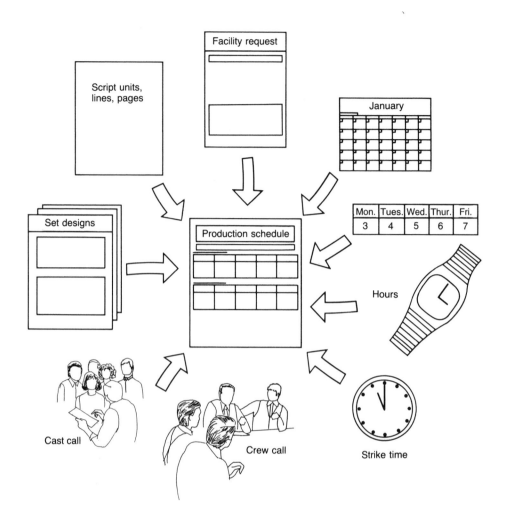

FIGURE 2–13
The production schedule. The producer creates the studio drama production schedule taking into consideration many factors.

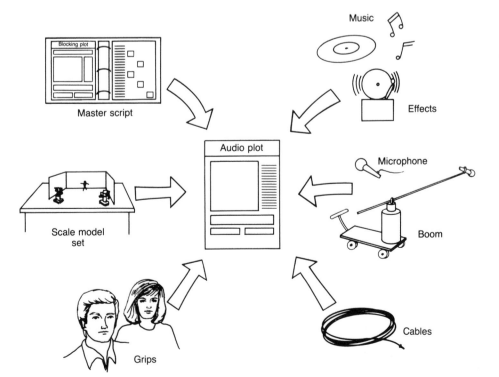

FIGURE 2–14
The audio plot. The audio director designs the audio plot weighing the elements of the master script and the requirements of the teleplay.

FIGURE 2–15
The audio design for the production. With the director's blocking plot, the audio director designs the microphone type, placement, and sound coverage for the set.

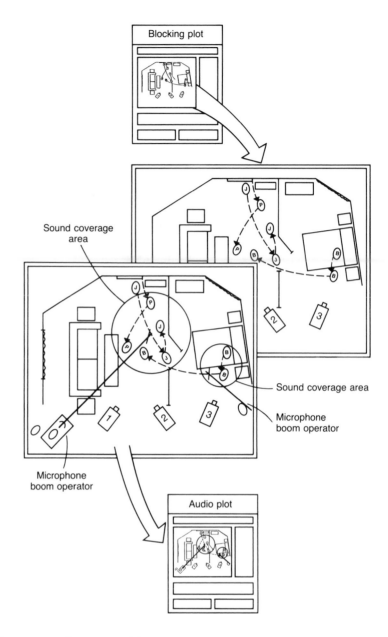

tions for grip(s)/operator(s) and microphone holder(s)/extension(s) during production. (See the audio pickup plot form.)

Preparing microphones, microphone holding hardware, and extensions (MBG/O 4) The microphone boom grip(s)/operator(s) prepare the microphones to be used in production and obtain and assemble the microphone holding hardware (e.g., fishpoles) and extensions.

Obtaining the necessary copyright clearances and royalty requirements (P 12) The producer is responsible for obtaining any copyright clearances and performance rights required for the production. Paramount among clearances are those for the music that the audio plot(s) may require. Obtaining clearances takes time, so as soon as the producer knows of required clearances the producer should begin procuring those clearances. The producer also has to settle any royalty rights accruing to

the owner of the copyright for the teleplay itself. Recording a copyrighted play on videotape requires a different type of clearance—performance rights—than the royalty for performing the play on the stage. In some instances, permission might be denied. Hence, the sooner performing rights for the play are requested and obtained the better for the production.

Securing insurance coverage (P 13) The producer is responsible for obtaining whatever insurance coverage the production may require. An example of required insurance coverage is protection coverage required by a facility if cigarette, cigar, or pipe smoking is required by the cast; if a real fire is required in a fireplace; or to cover the security of borrowed properties. Insurance may be required if small children not covered under any other insurance policies are part of a production and could be injured in the studio.

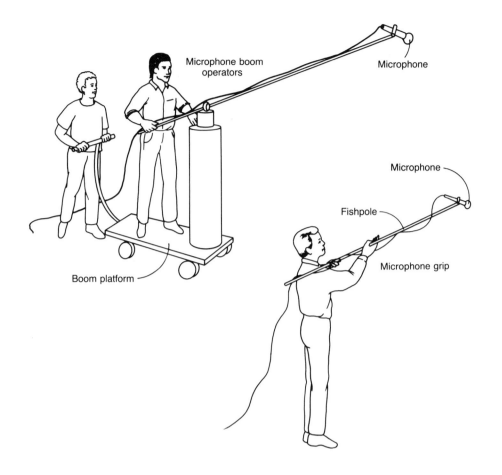

FIGURE 2–16
Microphone boom grip(s)/operator(s). Microphone boom grip(s)/operator(s) usually cover sound pickup with a platform microphone or a handheld fishpole.

Choosing and meeting with the lighting director and assigning the lighting plot (D 18) The director continues to build the production crew by bringing a lighting director into the crew. The director meets with the lighting director to convey the director's aesthetic design expectations for the teleplay. Lighting is a principal source for mood and environment and is critical in communicating some of the message of the production. The director gives the lighting director a photocopy of the master script for use in determining the set(s) and camera placement, location, time of day, set properties, and the blocking of the actor(s) or actress(es). The master script has all of these details.

Meeting with the director (LD 1) The lighting director meets with the director as a first step in designing lighting plot(s) for the production. The lighting director must have a copy of the master script as preparation to designing lighting for the production. The lighting director needs to know set design(s) including interior and exterior locations, time of day, set and action property placement, and actor and actress blocking with camera placement. The lighting director also needs to have the director's aesthetic sense for the mood to be conveyed with the production.

Working with the master script, set design(s), and properties list (LD 2) The lighting director must study the set design(s) and properties list and create breakdown notes as input in ultimately designing the lighting plot(s)

for the production. In some instances, knowing costume colors, make-up, hair color, and hairstyles is helpful to the lighting director in designing lighting.

Studying the scale model(s) of the set(s) (LD 3) The lighting director gains insight into lighting demands of the teleplay by studying the scale model(s) of the set(s). Proportional relationships and camera placement can be determined accurately with the scale model(s).

Designing the lighting plot(s) (LD 4) The lighting director designs the lighting plot(s) using the lighting plot form. (See the lighting plot form.)

Submitting the lighting designs to the director (LD 5) The lighting director reviews each set lighting design with the director. Final approval of any lighting design cannot come until the set(s) are complete and in the studio, lighting designs are in effect, actor(s) and actress(es) are in place with costumes and make-up, and the cameras are on.

Approving the lighting design(s) (D 19) The director reviews, makes any changes or suggestions, and approves the preliminary lighting design(s).

Approving the lighting design(s) and authorizing expenses and purchase requisitions (P 14) The producer reviews the lighting design(s), makes any changes deemed necessary, and approves the design(s). The pro-

FIGURE 2–17
The lighting plot. The lighting director must consider many elements of the teleplay and the production before designing the lighting plot(s) for the set(s).

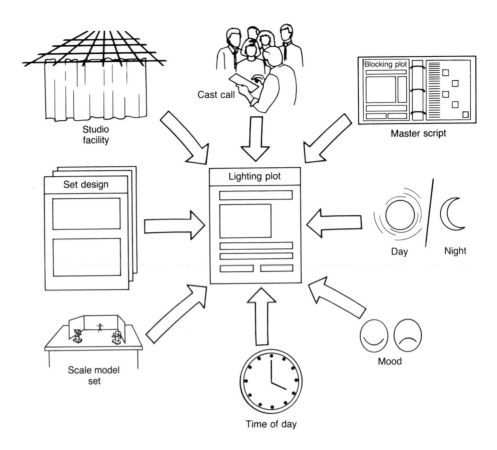

ducer also authorizes any expenses incurred and approves purchase requisitions for lighting needs.

Choosing and meeting with the technical director and assigning the shot list breakdown (D 20) The director chooses and meets with the technical director— an important co-director—as preproduction picks up for the teleplay. The meeting with the technical director covers the same explanation of expected given aesthetics to the other crew members and the plans for the production. The director gives the technical director a photocopy of the master script and a deadline for creating the shot list breakdown for the camera operators.

Meeting with the director (TD 1) The technical director meets with the director as a step toward beginning preproduction tasks. The technical director requires a photocopy of the master script in order to do a shot list breakdown for the camera operators.

Becoming familiar with the master script (TD 2) The technical director becomes familiar with the master script. The master script has all of the camera shots listed, which is the information the technical director requires in order to prepare the shot list breakdown.

Preparing the shot list breakdown from the master script (TD 3) The technical director prepares a shot list for each camera operator from the master script. The shot list is a listing of each planned camera shot in chronological order, but broken down for each camera operator. The shot list contains the number of the shot for each camera, the shot's assigned framing, the content and

composition of the shot, and any camera movement required. (See the camera shot list form.)

Designing titling and the credits list (P 15) The producer designs the order and content for all titling and credits for the production. This requires some aesthetic design because style and color are part of the choices available for character generator or computer designed graphics. A careful listing of all crew and cast plus support staff and their correct titles is part of the list of credits. (See the titling/credits copy form.)

Choosing and meeting with the assistant director and the production assistant (D 21) The director continues to round out the production crew by choosing an assistant director and the production assistant. The director then meets with both of them. The director informs the assistant director of the responsibilities to the production and to the director. Much of the role of the production assistant will take place during production. The director conveys the director's expectations to the production assistant. The character generator slate, title, and credits are the responsibility of the production assistant.

The assistant director has no preproduction responsibilities but to become familiar with the master script.

Choosing and meeting with the floor director and the continuity person (D 22) The floor director and the continuity person are chosen by the director. The floor director becomes the director's head and hands in the studio during production. The continuity person is responsible for observing and recording all production details of the set, props, dialogue, and actor(s)/actress(es)

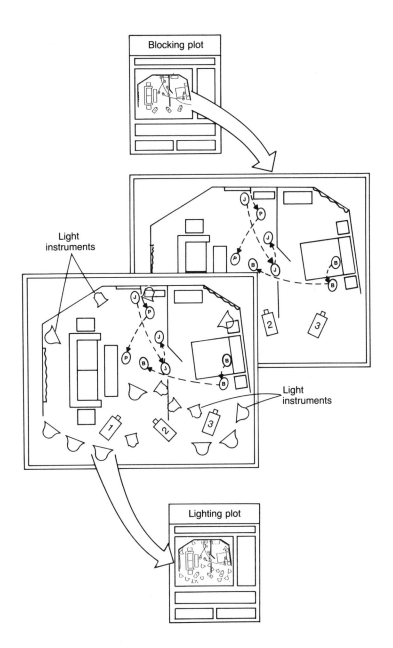

Blocking plot

Light instruments

Light instruments

Lighting plot

FIGURE 2–18
The lighting design. The lighting director designs the lighting for each set. Light instruments and their aim are drawn over a copy of the director's blocking plot for each set.

to reestablish them from the production of one script unit to another.

Meeting with the director (AD 1) The assistant director meets with the director. The assistant director should have a sense of how the director sees the assistant director's role during production. A principal component to assisting the director is to help control and organize the blocking and spiking of the actor(s), actress(es), and cameras. This is accomplished by using colored adhesive dots. The dots have to be purchased and a different color assigned to each actor, actress, and camera.

Meeting with the director (PA 1) The production assistant meets with the director and receives the expectations of the director for the production. The production assistant has important preproduction responsibilities, not the least of which is entering all of the screen text copy into the character generator. The character generator copy

includes the academy leader for every take, titling, and credits.

Doing preventative maintenance on the studio cameras (VEG 1) The video engineer spends preproduction time doing preventative maintenance on the studio camera. Preventative maintenance includes checking back focus on the cameras, checking for video image burn on the camera tubes, checking air pressure for camera pedestals, etc.

Routing the external signal to the camera monitors for camera operator use (VEG 2) The video engineer routes the external video signal from the line monitor in the control room to the monitors on the studio cameras. This will allow camera operators to cross check framing similarity with another camera when that camera is on line. This will permit aesthetically pleasing and balanced camera framing between actors in dialogue.

FIGURE 2–19
Camera shot lists. The technical director creates the camera shot lists by breaking down the three columns of storyboard frames (or camera shots) from the master script into consecutively numbered shots for each camera.

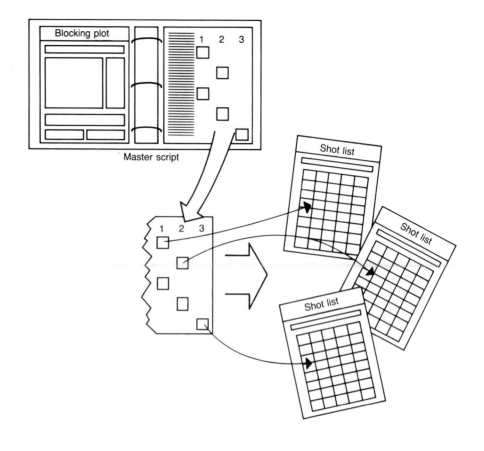

FIGURE 2–20
Titling and credits design. The producer is responsible for the accuracy, thoroughness, and design of the title screens and the complete cast and crew credits for the production.

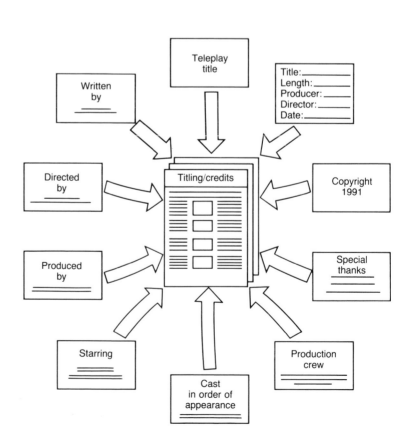

Securing adhesive colored dots and assigning a color to each actor, actress, and camera (AD 2) The assistant director secures or purchases enough and varied colored adhesive dots for each cast member and camera.

Assigning adhesive colored dots to each actor, actress, and camera (AD 3) The assistant director creates a master list of cast members on which to record the assigned color for each cast member and camera to be used during blocking and videotaping. The spike marks are placed on the studio floor by the assistant director during blocking and are removed after a studio wrap.

Securing titling and credits copy from the producer and entering the copy into the character generator (PA 2) The production assistant secures titling and credits copy from the producer, begins to enter the copy into the character generator, and records the information. The academy leader copy is also entered and recorded.

Meeting with the director (F 1) The floor director meets with the director as the first step in preparing for production. The floor director functions differently in studio drama production from the role of a floor director in other television genres. In studio drama production, the floor director functions as a stage director, talent call person, line prompter, and referee among conflicting artists. The director's expectations of the floor director are made clear during this meeting.

Studying the scale model(s) of the set(s) (F 2) A review of the scale model(s) of the set(s) can assist the floor director before getting to the studio. Studying the model(s) can help orient the floor director to the studio.

Meeting with the director (CP 1) The continuity person meets with the director to determine what expectations the director has for the observation and recording of production details.

Preparing forms for recording continuity details during production (CP 2) The continuity person prepares forms for recording sets, props, dialogue, and actor/actress continuity details during production. (See the continuity notes form.)

Choosing and meeting with the videotape recorder operator (D 23) The final production crew appointment is that of the videotape recorder operator. The director chooses and meets with the videotape recorder operator. The videotape recorder operator is responsible for securing adequate source videotape stock to videotape the entire teleplay and enough master tapes for postproduction editing. All of these tapes should be striped with SMPTE time code for ease in editing in postproduction.

Meeting with the director (VTRO 1) The videotape recorder operator meets with the director and receives preproduction assignments. The videotape recorder operator is responsible for securing an adequate quantity of videotape stock for videotaping the teleplay. Videotape stock should also be secured for postproduction editing of the master tape(s).

Securing the videotape stock for the production and postproduction (VTRO 2) The videotape recorder operator orders and purchases the videotape stock necessary to cover the production and postproduction editing needs.

SMPTE striping, coding, and labeling all videotape stock (VTRO 3) The videotape recorder operator stripes all videotape stock with SMPTE time code to make postproduction editing easier and faster. After striping, all videotapes should be coded for ease in accounting for a lot of source tapes later, and the videotapes and the videotape cases should be labeled accurately.

Completing set(s) construction and previewing the set(s) with the director (SD 8) The set designer oversees the completion of the set(s) required for the production and gives the director a preview of the completed set(s).

Previewing the completed set(s) with the set designer (D 24) The director previews the completed set(s) construction with the set designer, suggests changes, and gives final approval.

Lighting the studio set(s) before videotaping (LD 6) As soon as the studio set(s) are complete, the lighting director can begin placing lighting instruments, aiming them, and setting light intensity to create the lighting plot for each set.

Reviewing the set light pattern(s) with the director (LD 7) When the studio sets have been lighted following the approved lighting plots, the lighting director alerts the director and reviews the light patterns with the director.

Previewing preliminary lighting design on studio set(s) (D 25) The director previews, makes any changes, and approves the preliminary lighting designs on the studio set(s). This approval is on preliminary lighting because final lighting can be judged only with actor(s) and actress(es) in place and on-camera.

Reviewing and demonstrating acquired properties to the director (PM 7) When all properties are collected, the properties master/mistress reviews and demonstrates the properties for the director. The director needs a working knowledge of the handling and operation of all props. The operation of the properties must be approved by the director.

Previewing properties (D 26) The director previews demonstrations of acquired properties for the production. The director needs a working knowledge of how the properties function or handle before blocking actor(s) or actress(es). The director should make final changes or suggestions and approve the properties for the properties master/mistress.

Parading the final costumes before the director (CM 7) When costumes are completed, the costume master/mistress parades the costumes for the review of the director under studio lights and on-camera.

Parading in costumes for the director's preview and approval (T 5) The actor(s) and actress(es) are required to parade costumes for preview and approval or change by the director.

Reviewing costumes with the cast and costume master/mistress (D 27) The director previews a costume parade under studio lights and on-camera with the cast and costume master/mistress. The director makes changes or suggestions and approves the costume for production.

Testing make-up on actor(s) and actress(es) under studio lights and on-camera (MA 6) The make-up artist plans to see each actor and actress in full make-up under the studio lights and on-camera before final approval is sought from the director. This is an opportunity for the talent to learn their own make-up application and details before production.

Reviewing make-up designs and tests for the director (MA 7) The make-up artist reviews make-up designs under lights and on-camera for the director.

Trying make-up designs and testing the designs under lights and on-camera (T 6) Before studio production, the cast will have to try their own make-up designs on and test their make-up under studio lights and on-camera.

Previewing the make-up designs with the make-up artist under lights and on-camera (D 28) The director previews the video tests of all make-up designs under lights. The director can make suggestions or changes and then approve the tests.

THE PRODUCTION PROCESS

The production process for television studio drama contains elements quite unlike apparently similar elements in other studio genres. Studio drama is the most thoroughly—to minute details—preproduced video among television genres except, perhaps, for animation. A number of studio production roles differ in the extreme from the same roles in the other genres. The director directs not only from the control room, as in other genres but also in the studio. In other types of television studio production, the director has the freedom to choose any video source available in the control room during production, but in studio drama every camera shot is set, chosen, orchestrated, and preset before production begins. The technical director not only switches, as in any other television genre but also readies each defined and numbered camera shot over the intercom before it is taken. Camera operators not only operate their cameras but are blocked (positioned, assigned movement, and framed exactly) as are the actor(s) and actress(es) on the stage.

The cast experience is also different from that of the other television genres. Talent—actor(s) and actress(es)—function differently from the way they would in other television genres. Actor(s) and actress(es) come to television studio production without ever having been blocked for the set; their make-up, while generously applied, must be realistic looking—perfect—up close. Unlike proscenium theater, the talents' voices will be natural in tone and volume. The actor(s) and actress(es) will be required to go in and out of character many times and to hold reaction responses for unnaturally long periods. Movement for the actor or actress is muted, narrowly defined, restrained—cameras have to catch it. These and other differences make television studio drama production unique.

• Personnel

The production crew for television studio drama does not differ from the role labels of any other studio production. Definitions of those roles do differ, however. As in studio drama preproduction, the support staff is considered among the studio production personnel.

Producer (P) The producer's role during production is minimal, but supervisory. Working relationships—cast and crew relations and personnel and facility relations—occupy the producer more than anything else. The producer has the creative and aesthetic authority to accept or reject video takes during the production.

Director (D) The director is the heart and soul of television studio drama production—the center of a wheel from which all spokes emanate. In charge and in control, the director directs the cast and crew from the studio before videotaping and from the control room during videotaping.

Technical director (TD) A most important person in television studio drama production is the technical director. The technical director is the main point of communication during videotaping between the control room and the studio and between the director and the camera images.

Camera operators (CO) Another redefined role in studio drama versus that in the other television genres is the role played by the camera operators. The camera operator and camera are true players in the drama being produced. The camera and its operator have defined positions, spiked on the studio floor. The camera has exact framing and shot composition requirements. The movement of the camera is assigned, coordinated with talent lines or blocking. The camera and its operator are essentially considered as another actor or actress in the drama.

Lighting director (LD) Another artist in television studio drama is the lighting director. It is said that lighting makes television pictures possible. Lighting creates mood, atmosphere, and the environment necessary for drama. The lighting director cannot simply light a set and sit back and watch the teleplay. In television drama, shadows become a mar on the portrait. The lighting director is constantly vigilant during production to notice shadows and correct them.

Audio director (A) The audio director is challenged in the studio production of drama. While sound control is better in a studio, sound coverage is a challenge. Picking up talent without being seen either physically or by shadow is difficult. Essential sound elements to any drama are the artistic contributions of the audio director—music and sound effects.

Assistant director (AD) The assistant director's role in production is a support role. The assistant director works

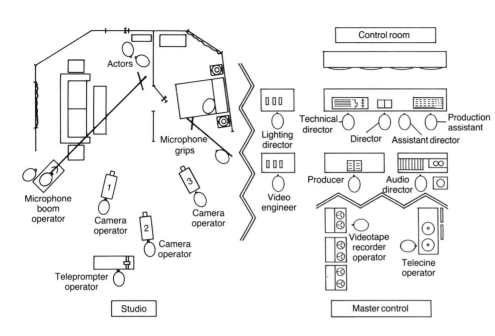

FIGURE 2–21
Production personnel placement. This diagram places each production personnel in the appropriate production facility area as required during videotaping.

closely with the director both in the studio, with actor(s) and actress(es) before videotaping, and in the control room during videotaping with the master script.

Production assistant (PA) The production assistant performs generally as the role might be defined in any other studio production—maintaining character generator copy and memory recall. The added requirement is the need for the slate before every take. The slate will also require constant update as videotaped takes progress and script units change.

Floor director (F) The floor director assumes more responsibility in drama production than the equivalent role in other studio productions. As the alter ego or other self of the director in the studio during videotaping, the floor director supervises the cast and support staff as well as the production crew. This often entails maintaining control between two sets of artists—the production and technical types and the creative and dramatic types. Once the director leaves the studio for the control room, the floor director will have to communicate more than production details to studio personnel: It often means interpreting the director's drama control to the actor(s), actress(es), and production crew.

Microphone boom grip(s)/operator(s) (MBG/O) Additional crew to studio production not found in the other television genres are the microphone boom grip(s)/operator(s). With the dynamism of extensive movement of the actor(s) and actress(es) in large sets, the task of sound coverage is compounded. Microphone boom operator(s) are necessary if the studio uses a platform boom; microphone grip(s) are necessary if the sound coverage will be accomplished by handheld extension boom(s) or fishpole(s). The compounding challenge to the microphone boom grip(s)/operator(s) is to keep themselves, their hardware, and their shadows from appearing on-camera.

Videotape recorder operator (VTRO) The videotape recorder operator has the production responsibility for acquiring the videotape stock necessary for the production

and properly labeling and coding each new videotape and case. The videotape recorder operator will be constantly needed at the videotape recorder position for the many short videotaped units required during teleplay production.

Video engineer (VEG) The video engineer is responsible for the continued monitoring of the studio cameras and the operation of studio floor monitors. Monitoring cameras includes watching video light levels over waveform oscilloscopes as cameras and talent move about the studio sets and preparing to adjust lens aperture accordingly. The video engineer is challenged during teleplay production by the more dynamic movement of multiple studio cameras under creative lighting design.

Talent/actor/actress (T) The talent—actor(s) and actress(es)—take on different requirements in television studio drama production than in the other television genres. They meet at cast call without having been blocked. Actor(s) and actress(es) come to the studio as other than the center of attention; in studio drama production, the process of videotaping is the center of attention. In the studio, hardware—studio cameras—become players with the actor(s) and actress(es) in production. The big challenge to the actor(s) and actress(es) in studio drama production is to maintain character in the midst of what will appear to be technological confusion. Maintaining character will be further challenged by the constant take and retake of script units. Another challenge is the requirement to ''hurry up and wait.'' Being around but out of the way will test the patience of both the cast and crew. Needing to be elsewhere to change or correct costuming, make-up, or properties and still be ready to be blocked or videotaped challenge both cast and crew, as will picking up lines, blocking, and characterization in the middle of dialogue.

Costume master/mistress (CM) The costume master/mistress has work that is well defined. Overseeing the costumes and accessories, dressing actor(s) and actress(es), cleaning and repairing costumes, tracking accessories, and checking costume detail keep the costume master/mistress occupied throughout production.

Properties master/mistress (PM) The properties master/mistress has much to do during production. Set, action, and hand properties are the constant responsibility of the properties master/mistress. The only time props are not the responsibility of the properties master/mistress is while they are being used in production. Immediately before and after use, the props are in the care and security of the properties master/mistress. An ordered storage area in the studio near the set(s), a place assigned for each prop, and a checklist of every property required become the production tasks for the properties master/mistress.

Make-up artist (MA) The make-up artist has production responsibilities that do not end until the talent have left the studio. Generally, talent do not get into make-up until after blocking and rehearsal (i.e., during the break between the rehearsal and the first videotape take). Make-up that requires extensive application time (e.g., for aging or a deformity) can begin before blocking. While most actors and actresses may apply their own make-up after testing during preproduction, the make-up artist is still needed for touch ups and details. In the studio, the make-up artist is constantly needed for make-up repairs.

Set designer (SD) The set designer is not usually required to be in the studio during production, but, having designed the set(s) for the teleplay, the set designer may choose to be in the studio at least for the first videotaping use of each set. This gives the set designer the opportunity to dress the set according to the approved designs for the first time each set is used.

• **Production Stages**

Supervising cast and crew calls (P 1) The producer begins the studio production process by supervising the scheduled cast and crew calls. The crew call (and cast call) is a scheduled gathering or rendezvous time for production personnel. Supervisory personnel often use the crew call as a time for a production meeting also. The cast and crew calls may be a joint meeting time or separate. Some actor(s) and actress(es) may need a lot of lead time in order to be ready (e.g., to allow for detailed make-up preparation) for production. This may entail an earlier cast call. On the other hand, the production crew needs enough lead time to be at a point of readiness when the talent arrive. Studio production is often characterized as a "hurry up and wait" phenomenon. Adequate crew call lead time can remedy that problem for the cast. This time is an opportunity for the producer and director to convey last minute details or changes in production that pertain to the crew and cast.

Meeting cast and crew calls and holding a production meeting (D 1) The director begins studio production by meeting with the cast and crew at either joint or separate calls. The director uses this rendezvous time as the opportunity to hold a production meeting to set expectations for the day's production, announce changes to the master script or to any other production element, announce crew or facility changes, and field any questions

the crew or cast may have concerning the production session. The director then reviews the production schedule for the remaining production sessions.

The director concludes the production meeting by setting a time convenient to the crew for set-up and the cast for costume and make-up preparation for the first blocking of the production day. Apart from extensive make-up application, only the production crew need have their equipment ready for blocking.

Handling studio facility and set(s) arrangements (P 2) The producer follows through with the role responsibility for the producer from preproduction and concludes by checking for expected facility arrangements determined and ordered during preproduction. This includes checking on everything from air conditioning in the studio to dressing on the set(s). The producer initiates facility relations before production and maintains the relationship with facility supervisory personnel during production. Any additional requests and changes regarding the production facility should go through the producer.

Overseeing the decoration of the set(s) before production (SD 1) The set designer oversees the details of set decoration for the set(s) to be used in the particular production session. There is no better resource for proper display and arrangement than to have the designer of the set make the final touches before production.

Supervising final studio facility and set(s) arrangements (D 2) The director walks through the studio production facility and the studio set(s) as a final detail check before and as the crew begins to set up for production. This permits the director the opportunity to make the producer aware of anything missing, not working, or not yet in place for production.

Readying the cameras for shading and videotaping (VEG 1) As support for the studio set-up, the video engineer readies the cameras by first electronically uncapping the cameras. The video engineer helps the camera operators to prepare their cameras. The video engineer has engineering responsibility over the cameras whereas the camera operators have operation responsibility.

Handling production crew and cast details (P 3) The producer handles personal details for the members of the cast and crew. This can entail anything from arranging for storage for securities during production to providing hot water in the make-up room, coffee in the green room, or parking for personal cars.

Supervising equipment set-up and placement (D 3) The director oversees all equipment set-up for the production. The sooner the director gets around crew positions and completes the checks on hardware and operational conditions the sooner the director can begin blocking. The breadth of personnel and equipment that has to converge into the moment of beginning videotaping is so extensive that the director has to get on top of missing equipment, inoperative hardware, or other difficulties, so the producer can trouble-shoot and solve the problems. Crew members need to be encouraged to bring to the attention of the director and producer as soon as possible any difficulties in production elements.

Meeting with the producer and director for cast call and the production meeting (T 1) The actor(s) and actress(es) meet for the scheduled cast call and at the production meeting with the producer and director.

Meeting with the producer and director for cast call and the production meeting (PM 1) The properties master/mistress meet with the cast for cast call and with the producer and director for the production meeting.

Meeting with the producer and director for cast call and the production meeting (CM 1) The costume master/mistress meets with the cast for the cast call and with the producer and director for the production meeting.

Meeting with the producer and director for cast call and the production meeting (MA 1) The make-up artist meets with the cast for the cast call and with the producer and director for the production meeting.

Completing set(s) decoration with set, hand, and action properties (PM 2) The properties master/mistress completes set(s) decoration with the set, hand, and action properties needed for the day's production. Because most properties are not left lying around before and after production, properties must be gathered from storage and security, laid out, and placed on the set(s) before each production session. Many consumable props (e.g., food, beverages, and cigarettes) have to be prepared or replenished.

Meeting with the producer and director for crew call and production meeting (AD 1) The assistant director meets for crew call and production meeting with the producer and the director.

Meeting with the producer and director for crew call and the production meeting (CO 1) The camera operators meet for crew call and the production meeting with the producer and director.

Meeting with the producer and director for crew call and the production meeting (TD 1) The technical director meets with the producer and director for crew call and the production meeting.

Meeting with the producer and director for crew call and the production meeting. (A 1) The audio director meets with the producer and director for crew call and the production meeting.

Meeting with the producer and director for crew call and the production meeting (MBG/O 1) The microphone boom grip(s)/operator(s) meet with the producer and director for crew call and the production meeting.

Meeting with the producer and director for crew call and the production meeting (F 1) The floor director meets with the producer and director for crew call and the production meeting.

Meeting with the producer and director for crew call and the production meeting (LD 1) The lighting director meets with the producer and director for crew call and the production meeting.

Meeting with the producer and director for crew call and the production meeting (VTRO 1) The videotape recorder operator meets with the producer and director for crew call and the production meeting.

Meeting with the producer and director for crew call and the production meeting (PA 1) The production assistant meets with the producer and director for crew call and the production meeting.

Meeting with the producer and director for crew call and the production meeting (CP 1) The continuity person meets with the producer and director for crew call and the production meeting.

Checking video levels of cameras with the lighting on the set(s) (VEG 2) The video engineer requires that the camera operators aim and focus their cameras on the same object within the lighted set. This allows the video engineer to shade and set lens levels for optimum videotaping. Shading requires setting contrast ratios between light and dark areas of the lighted set.

Alerting camera operators and the director when ready (VEG 3) The video engineer alerts the camera operators and the director when the cameras are shaded and ready for videotaping. The cameras may now be moved from their aimed and focused position.

Setting up and preparing assigned cameras and placing them for first blocking (CO 2) The camera operators set up their cameras and prepare them for first blocking. Preparation involves uncapping the lenses at the camera, adjusting pan and friction tension, setting or balancing the pedestal operation, adjusting control and zoom arms to a comfortable height, placing each camera in its proper order from left to right (camera 1, camera 2, camera 3), and drawing sufficient camera cable behind each camera to allow for adequate movement.

Reviewing the shot list and attaching the shot list to the back of the camera (CO 3) The camera operators review the shot lists prepared by the technical director and distributed during preproduction. Each shot list, listed by camera, contains all of the prepared shots for each camera from the director's master script. While these lists are subject to change during blocking and rehearsal, they are a good point of reference for the camera operators. The more familiar the camera operators are with the demands to be made on the camera and its operator, the easier will be the blocking and rehearsal for the camera operators. Each shot list is attached behind the camera to which it applies and below the camera monitor.

Lighting the set(s) and performing the lighting design check (LD 2) The lighting director lights the set(s) with the lighting design that was completed and approved during preproduction. Given some passage of time from lighting design completion to the production session, the lighting director must check carefully that all prehung and preset lighting instruments are in place, aimed correctly, and at the preset intensity. The lighting director may call the cast or crew to stand-in for an actor or actress while the lighting is checked.

Meeting with the microphone boom grip(s)/operator(s) and setting up the audio equipment (A 2) The audio director meets with the microphone boom grip(s)/operator(s) and begins setting up the audio equipment for production.

Meeting with the audio director and beginning audio equipment set-up (MBG/O 2) After the production meeting, the microphone boom grip(s)/operator(s) meet with the audio director to begin setting up the audio equipment for production.

Assembling microphone booms with microphones and running audio cables (MBG/O 3) The microphone boom grip(s)/operator(s) begin assembling whatever audio equipment remains to be assembled from preproduction. Microphone cables have to be run from the set(s) and connected to the wall or box connector.

Running microphone cables and recording studio microphone inputs (A 3) As the microphone boom grip(s)/operator(s) complete cable runs, the audio director records the wall or box input connections in the studio for the microphones.

Patching studio microphone inputs into the audio control board in the control room (A 4) The audio director patches the corresponding input selection for microphones from the studio into the audio board in the control room.

Making intercom connections (A 5) The audio director must select those intercom connections needed during production. The audio director will need principal communication with the technical director and the director (if the director chooses to be on the intercom).

Preparing and assisting the director with the master script (AD 2) The assistant director prepares to assist the director. Primary assistance is with the master script. The assistant director needs to have the director's copy of the master script and to follow the director closely during the preblocking stages of production.

Assuming responsibility for the studio, set(s), cast, and crew during studio use (F 2) As studio set-up progresses, the cast prepares and other principals (e.g., set designer and properties master/mistress) relinquish their primary obligations while the floor director begins taking control and responsibility in the studio. The floor director's control begins with the awareness of the state of preparation of all elements of the studio. The floor director wants to be able to report studio readiness to the director before being asked or certainly when asked.

Checking intercom connections (F 3) The floor director checks the headset and intercom connections with the technical director and the director (if the director chooses to be on the intercom network).

Checking intercom connections (TD 2) The technical director checks the intercom connections with the camera operators, floor director, audio director, production assistant, videotape recorder operator, and director (if the director chooses to be on the network).

Checking intercom connections (PA 2) The production assistant checks the intercom connections needed at the character generator. The production assistant needs contact with the technical director and the director (if the director chooses to be on the network).

Knowing what the director needs and requires during rehearsal(s) and take(s) (F 4) Important to the production process and the role of the floor director is to know at all times what the director needs and requires during all stages of production. This means the knowledge of where the director intends to begin with the script, any changes in production order, and the cast required for any stage. Much of this information is contained in the production schedule form that was prepared during preproduction and reviewed at the production meeting.

Entering and recording character generator copy for the script units to be produced (PA 3) The production assistant completes whatever character generator copy was not entered and recorded during preproduction. The production assistant prepares for last minute needs of the character generator for the script units about to be produced. This preparation includes not only the updated slate for the academy leader but also any titling or credits to be put on videotape during any script unit to be produced. This information can be found from the master script and from the production schedule distributed during preproduction and updated during the production meeting.

Preparing the record videotape deck and selecting labeled videotape stock (VTRO 2) The videotape recorder operator prepares the record videotape deck for recording production takes by selecting the appropriately labeled videotape stock assigned for the production.

Readying forms for recording continuity notes during rehearsals and takes (CP 2) The continuity person readies continuity recording forms for the script units scheduled for the production session. Readying the continuity forms requires labeling the individual forms for the specific units to be blocked, rehearsed, and videotaped. Relevant information for the continuity forms is available from the production schedule form and from the videotape recorder operator. The videotape recorder operator has the code(s) for the videotape stock and source tape(s) being used for the script units recorded during the production session.

Meeting the lighting director's requirement for lighting stand-in needs (T 2) The actor(s) and actress(es) meet the lighting director's requirements for standing in for the final aiming and intensity setting of lighting instruments during preparation for blocking the set(s). The person to be lighted is required as a stand-in for final adjustments in lighting in order to adjust for such attributes as height, coloring, and clothes.

Checking intercom connections (VTRO 3) The videotape recorder operator checks the required intercom connections. The videotape recorder operator needs contact with the technical director and the director (if the director chooses to be on the network).

Checking intercom connection and announcing readiness (CO 4)   The camera operators check that they have the correct intercom connection (i.e., with the technical director) and then alert the floor director that they are ready.

Checking the intercom connections for the director (AD 3)   The assistant director checks any intercom connections for the director. Some directors may not want to be on the intercom network during production because they do not have to be.

During production, most directors prefer to concentrate on the camera monitors, the master script, and the choice of cutting as planned. The director is adjacent to the technical director and, during production, only the technical director needs to hear the director. The director may want intercom connections to the technical director and the videotape recorder operator. These connections should be selected and checked.

Calling and readying actor(s), actress(es), and cameras by script unit and spiking actor(s), actress(es), and cameras (AD 4)   The assistant director calls and readies the cast and crew to the beginning of blocking. This call should be a reminder as to the exact script unit being blocked. The assistant director readies the adhesive colored dots for spiking the actor(s), actress(es), and cameras during blocking. The adhesive dots are color coded for each actor, actress, and camera.

Determining B-roll videotape needs from the master script (VTRO 4)   The videotape operator can determine from the master script when videotaped takes will be required for postproduction techniques such as dissolving or special effects. When takes will be dissolved or used

with special effects, one video source should be on an A-roll and the second source on a B-roll. When these are anticipated and prepared during preproduction, takes can be recorded on separate videotapes (one videotape for A-roll, and a second videotape for B-roll) for postproduction use. This saves video signal loss (a second generation) when a duplication of a take has to be made from a single videotape.

Following the director's directions for blocking; rehearsal(s); videotape take(s); and calls for action, cut, or freeze (T 3)   The cast should be prepared from this point on to follow without comment any directions that the director may give to the cast. Those directions include the command to begin the action of the script unit, to cut or stop at any point in a unit, or to freeze where they are. The freeze is called to enable the continuity person to record details of costuming, movement, and property use and to facilitate editing any subsequent take(s).

Following the master script with and for the director (AD 5)   The assistant director handles the master script and follows it for the director as the assistant director follows the director during blocking. Any changes the director may make during blocking and rehearsal should be recorded in the director's master script.

Following the director with a copy of the master script that will be used to call shots over the intercom during later rehearsal(s) and take(s) from the control room (TD 3)   The technical director follows the director through blocking and rehearsal(s) with the technical director's copy of the master script. This is the script the technical director will use from the control room during later rehearsal(s) and take(s) to ready shots. Any changes

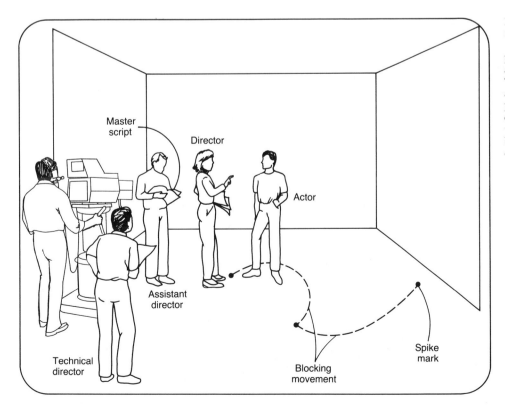

FIGURE 2–22
Studio drama production: Step 1. The first step in studio drama production is the blocking of the actor(s)/actress(es) by the director. The director physically moves the actor(s)/actress(es) from point to point. The assistant director spikes each rest point of an actor or actress with a colored adhesive dot. This blocking is done without the actor or actress delivering lines of script.

the director may make will have to be made during blocking in the technical director's script copy.

Beginning blocking and blocking actor(s)/actress(es) without lines (D 4) When the appointed time for beginning blocking arrives, the cast and studio personnel should be ready. The assistant director and technical director accompany the director on the set scheduled for the first script unit. Camera operators, audio director, and microphone boom grip(s)/operator(s) should be ready to observe blocking.

The director begins blocking, which, for the director, is the first step in actual production. This is the first interface between the cast and crew for the purpose of videotaping. The cast and crew are aware from the production schedule form what script units, set, cast, properties, and costumes are required as well as the studio blocking starting time. At this point, the cast is not in costume or make-up. Most drama production processes take a break between the last rehearsal after blocking and the first taping for make-up and costuming. If the length of time needed for applying make-up or dressing is excessive, make-up and costuming may have been begun and even been completed before blocking. This is an exception, though.

The director blocks actor(s) and actress(es) by physically moving each, one at a time, from a starting position within the set through a movement to a rest position, line by line as preplanned from the director's master script. At each rest position from start to finish, the assistant director places an adhesive dot between the feet of each of the actors and actresses as they are blocked. The dots are color coded for each cast member.

Updating the technical director's master script copy (TD 4) As the technical director follows the director through the blocking of the actor(s), actress(es), and cameras, any changes (additions, eliminations, or alterations) to the original master script have to be made in the technical director's copy. The technical director's copy of the master script will be used during production from the control room to ready all camera shots. It must be accurately updated.

Rehearsing the blocking of the actor(s)/actress(es) without their lines (D 5) The director now rehearses the actor(s) and actress(es) through the blocking *without* the actor(s) and actress(es) delivering their lines. The director watches the cast members hit their assigned spike marks (colored dots).

Rehearsing the actor(s)/actress(es) with their lines (D 6) The director now rehearses the actor(s)/actress(es) while delivering their lines. The director watches and affirms or corrects the rehearsal. The director may rehearse a second time with lines before videotaping.

Standing by to handle action properties before and after all rehearsal(s) and take(s) (PM 3) The properties master/mistress stands by the set during all rehearsal(s) and take(s) to handle and store all action props used by the actor(s) and actress(es). Many props will have to be replenished or reloaded, depending on the nature of the property. This is the responsibility of the properties master/mistress.

Observing the blocking and rehearsal of the actor(s)/ actress(es) by the director (CO 5) The camera operators

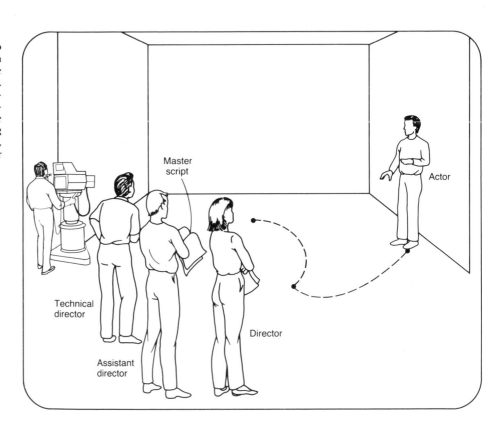

FIGURE 2–23
Studio drama production: Step 2. The second step in studio drama production is a rehearsal of the blocking by the actor(s)/actress(es). First, the actor(s)/actress(es) rehearse the blocking without delivering their lines. They then rehearse the same blocking while delivering their lines. The director watches, corrects, changes, and calls another rehearsal.

Master script

Actor

Technical director

Assistant director

Director

observe the details of the blocking and rehearsals of the actor(s) and actress(es). This orients the camera operators to what will be expected of them.

Placing microphone boom grip(s)/operator(s) and microphone boom(s) for the script unit being blocked (A 6) The audio director carefully watches the blocking together with the microphone boom grip(s)/operator(s) for planning placement of the grip(s) and boom(s) to cover studio sound.

Taking place(s) for preliminary dialogue and sound coverage (MBG/O 4) The microphone boom grip(s)/operator(s) take their preplanned places in the set for preliminary dialogue and studio sound coverage. This step for the audio director is just a beginning attempt to find and test the best places in the set for sound coverage.

Blocking the cameras as actor(s) and actress(es) again hit their spike marks (D 7) The director blocks each camera physically as the actor(s) and actress(es) hit their spike marks again, usually without their lines, or, certainly, by delivering their lines slowly.

The director takes each successive camera and its defined placement and has the assistant director spike the camera's placement with the colored adhesive dots. The spike marks for cameras are usually placed directly under the center of the camera's pedestal trunk. Then the director sets the defined composition (e.g., actor or actress) and its framing (e.g., CU or LS) for each shot. Any secondary movement (e.g., zooming or dollying) is blocked. All of these directions should correspond to the preproduction shot lists the camera operators received from the technical director. The camera operators should update their own shot lists, making any additional memory notes on their shot lists to assist them in getting each shot during rehearsal(s) and take(s).

Being blocked with the actor(s) and actress(es) and spiking the cameras (CO 6) The cameras are now blocked just as the actor(s) and actress(es) were blocked. The camera operators should see their cameras and themselves as other actor(s)/actress(es) in the script unit. They will have to be in their exactly defined places with the assigned camera composition and framing as the actor(s) and actress(es) will be required to be on their spike marks when lines are delivered and movements are made. The assistant director spikes each camera placement with an assigned colored adhesive dot.

Noting details of camera placement, shot composition, framing, and making changes on the shot list (CO 7) Each camera operator makes notes on the shot

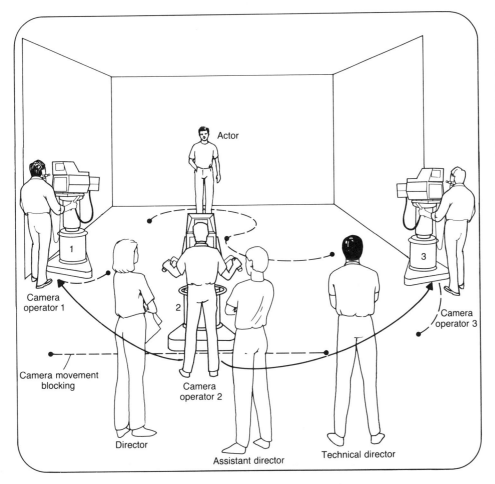

FIGURE 2–24
Studio drama production: Step 3. The director blocks each studio camera, one at a time, in relation to the previous blocking of the actor(s)/actress(es). Every camera position is spiked. The actor(s)/actress(es) perform their blocking without delivering their lines while blocking of the cameras continues. The director sets framing and content of each camera lens as part of blocking.

Actor

1

3

2

Camera operator 1

Camera movement blocking

Director

Camera operator 2

Assistant director

Technical director

Camera operator 3

list for the camera of any details of the placement of the camera, the shot composition, and framing. Any changes and additions to the shot list should be made as each shot is assigned by the director. Camera operators should also make their own notes for each shot to help them get the required shots again during rehearsal(s) and take(s).

Double checking all changes to the master script on each camera's shot list (TD 5)   After the director finishes with each camera, the technical director follows and checks each camera shot list with the master script as a check that all changes were made and noted on the shot list.

Rehearsing cameras, actor(s), and actress(es) (D 8)
The director rehearses the cameras and cast for the first time by loudly calling out in advance the number of each defined shot. (The assistant director may be asked to call the shot numbers for the director.) This may be the most tedious step in the production process as cast and crew attempt to integrate and coordinate the blocking of each. The cast may be asked to slow their pace of delivery and movement for this step.

As this rehearsal progresses, the audio director and the microphone boom grip(s)/operator(s) also attempt to integrate their studio sound coverage. They begin by taking their assigned places in the set.

The lighting director begins to monitor lighting from the shots viewed over the camera monitors. The director may move from camera monitor to camera monitor to check the defined shots. Because the technical director is still in the studio with the director, there is no way to cut

shots at the switcher to allow shots to be monitored from the studio floor monitor.

The director relies on feedback from the camera operators regarding individual defined shots and on the operator's ability to get them as designed.

Rehearsing cameras with actor(s) and actress(es) (CO 8)   The camera operators rehearse their defined shots as the actor(s) and actress(es) rehearse their dialogue and movement. The camera operators need to follow their shot lists scrupulously. This is the time to discover any difficulties in getting assigned shots, moving the cameras, framing the shots, and focusing the lens.

Making changes in blocking and continuing rehearsal in the studio (D 9)   The director listens to the camera operators and cast on the difficulties they may have in achieving their defined blocking. Then, by trouble-shooting the difficulties and retaining the original blocking or by changing the blocking, the director rehearses from the studio again. The director continues these steps until the cast and crew achieve the required shots.

Updating camera, actor(s), and actress(es) changes on the shot list (CO 9)   The camera operators update and make any changes to their camera blocking, numbering of shots, and actor or actress information on their shot lists. The camera operators inform the director of all difficulties they experience in getting the assigned shots and make changes accordingly as the director instructs them.

FIGURE 2–25
Studio drama production: Step 4. In this step, the director rehearses the actor(s)/actress(es) and cameras. Microphone grip(s) and operator(s) are added. The director monitors this rehearsal from each camera's monitor as the assistant director calls each shot out loud from the studio.

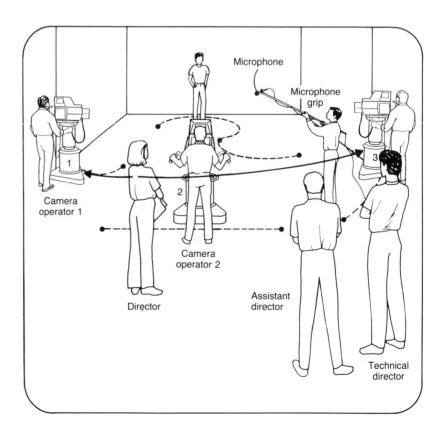

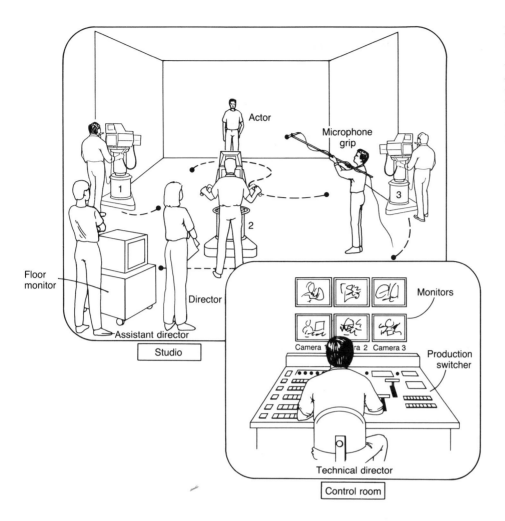

FIGURE 2–26
Studio drama production: Step 5. For this step, the technical director retires to the control room and calls and cuts each shot from the switcher. The director watches the rehearsal on the floor monitor in the studio. Cameras, actor(s)/actress(es), and microphone grip(s) all rehearse their blocking with lines.

Rehearsing as often as the director calls for rehearsals (CO 10) The camera operators rehearse as often as the director calls for rehearsals, getting all assigned shots and details of shots as directed.

Retiring to the control room and switcher and responding to the director's calls from the studio (TD 6) When the director feels it is convenient, the technical director retires to the control room and takes the technical director's place at the switcher. The director calls the shots from the studio over either the microphones or a head-set. The technical director then responds to the director's calls from the studio and cuts the camera shots as called by number. This allows the director to watch the studio floor monitor for the succession of camera shots as designed. In time, the technical director can begin readying the shots and cutting them for the director while the director is still in the studio.

Updating the master script copy with all changes made during rehearsal(s) by the director (TD 7) The technical director updates the technical director's copy of the master script with all changes made by the director during rehearsal(s). This is the script from which the technical director will make camera calls during videotaping.

Watching camera and cast blocking for microphone sound coverage and conferring with the audio director (MBG/O 5) The microphone boom grip(s)/operator(s) watch the camera and cast blocking and rehearsal(s) for microphone sound coverage possibilities. The grip(s)/operator(s) confer with the audio director as to the best microphone placement and try sound coverage during rehearsal(s).

Beginning microphone sound coverage during camera, actor(s), and actress(es) rehearsal(s) with lines (MBG/O 6) The microphone boom grip(s)/operator(s) begin sound coverage as soon as the director rehearses the actor(s) and actress(es) with lines. The best thing the grip(s)/operator(s) can do is to attempt sound coverage during this rehearsal. Given the many factors and difficulties of sound coverage, the sooner they attempt to cover dialogue the sooner they can discover problems and work them out. In addition to providing sound coverage, the grip(s)/operator(s) have to avoid getting on-camera, catching their microphone(s) on-camera, or casting shadows on the set. Problems of coverage have to be identified and solved as soon as possible.

Conferring with the audio director for improved microphone placement and sound coverage and mak-

ing changes to avoid getting on-camera or casting shadows (MBG/O 7) The microphone boom grip(s)/operator(s) confer continuously with the audio director for improved microphone placement and sound coverage during all rehearsals. The grip(s)/operator(s) should make those changes that get them and their equipment off-camera and eliminate any shadows.

Noting CU/LS camera framing for improved sound perspective placement of microphone(s) (MBG/O 8) The microphone boom grip(s)/operator(s) note when cameras have close-up or long shot framing assigned to them. This allows the sound coverage to begin to achieve better sound perspective. When a shot is a close-up, the audience should perceive dialogue sound closer. This can be achieved by placing the microphone closer to the actor or actress speaking. Because the shot is framed tighter, the microphone can safely be placed closer without its being seen on-camera.

If a camera shot of an actor or actress is a long shot, the audience expects the dialogue to sound more distant. Because the shot is framed longer, the microphone has to be placed farther from the actor or actress to avoid being caught in the long shot. This should give better sound perspective to the audience. Good grip(s)/operator(s) begin to memorize dialogue and know camera shots during rehearsal(s) as part of their better sound coverage techniques.

Making microphone audio levels checks as needed (A 7) The audio director makes microphone sound level checks (e.g., volume, equalization, and tone) during rehearsal(s) as often as possible to achieve acceptable levels. Difficulties in achieving acceptable levels require trouble-shooting from the studio with the microphone boom grip(s)/operator(s).

Making studio foldback sound checks for microphone boom grip(s)/operator(s) (A 8) The audio director patches and checks studio foldback sound for the microphone boom grip(s)/operator(s). The studio grip(s)/operator(s) need to monitor the dialogue over headsets for sound coverage during production.

Monitoring cast dialogue with foldback over headsets during rehearsal(s) and take(s) (MBG/O 9) Microphone boom grip(s)/operators(s) monitor their own sound coverage pickup by audio foldback over headsets. This permits them a better sense of their own sound coverage and microphone placement during rehearsal(s) and take(s).

Rehearsing sound effects into the studio (A 9) The audio director interfaces any required sound effects (e.g., telephone or doorbell ringing) into the studio rehearsal(s) when the director calls for them to be integrated. Speakers in the studio have to be checked for sound reproduction.

Watching for microphone(s), boom(s), grip(s), and shadows on shots during rehearsal(s) (CO 11) The camera operators watch their camera monitors closely for signs of any microphones, boom(s), grip(s) or their shad- ows in any shot. The audio director should be alerted if any of these appear in the shots.

Observing camera and cast blocking and directing microphone boom and grip placement for better sound coverage (A 10) The audio director constantly observes camera and cast blocking and directs microphone boom and grip placement for better sound coverage of dialogue from the studio. The audio director observes camera and cast blocking on the audio control board monitors or from the studio.

Watching set(s), actor(s), and actress(es) for light and shadows (LD 3) The lighting director watches from the studio and on the control room monitors for any problems in lighting the set(s) or the cast during rehearsal(s). The lighting director can anticipate lighting needs to a point. Once blocking occurs and rehearsal(s) begin, problems of overlighting, underlighting, or excessive reflection and glare can appear.

Adjusting lighting instruments or lighting intensity after each rehearsal (LD 4) The lighting director makes any changes in lighting instruments or lighting intensity between rehearsals.

Rehearsing with actor(s), actress(es), and cameras (MBG/O 10) The microphone boom grip(s)/operator(s) begin rehearsing their microphone placement and sound coverage as often as the cast and cameras rehearse. It is always more convenient to integrate sound coverage as others rehearse rather than needing a separate rehearsal for sound coverage alone. Many rehearsals may be needed to trouble-shoot the problems possible with sound coverage.

Making necessary changes and adapting (MBG/O 11) The microphone boom grip(s)/operator(s) make necessary changes in sound coverage as problems are solved and adapt microphone placement and sound coverage techniques whenever necessary. Grip(s)/operator(s) should consider all possible microphone placement necessary to cover sound (e.g., over the top of a set).

Monitoring microphone placement and sound levels during rehearsal(s) (A 11) The audio director is responsible for microphone placement and monitors the positions of the microphone boom grip(s)/operator(s) during rehearsal(s). The audio director constantly monitors and checks sound levels during rehearsals.

Making microphone grip(s)/operator(s) placement changes before any successive rehearsal (A 12) The audio director has the final authority for the placement of grip(s)/operator(s) and microphone(s) during rehearsals and in preparation for videotaping.

Mixing in any sound or music effects as required by the script (A 13) The audio director mixes in any sound effects or music as required by the script. Audio levels have to be preset for these audio sources.

Monitoring the dialogue and sound effects during rehearsal(s) (A 14) The audio director monitors all audio sources during rehearsal(s). This includes the dialogue and sound effects.

Retiring to the control room for a rehearsal (D 10) When the director feels that there have been sufficient rehearsals and all necessary changes integrated, the director retires to the control room with the assistant director. The technical director is already in the control room. The director then proceeds with a rehearsal from the control room. This means that the director will follow the master script, assisted by the assistant director, and will call to the technical director to take the defined camera shots when they are to be taken. If other changes to cast and/or crew are to be made, the director makes them over the intercom network to the studio.

Studying lighting effects over the floor monitors and making final changes (LD 5) The lighting director takes the opportunities afforded by rehearsals to study the lighting design effects over the studio monitors as the actor(s), actress(es), and cameras move about in the set. Any undesirable effects are changed between rehearsals.

Making changes and updating the master script for the director (AD 6) The assistant director makes any changes in cameras, shot numbering, or cast blocking on the director's master script.

Preparing for defined camera shots by number to camera operators over the intercom during rehearsal(s) when the director retires to the control room (TD 8) The technical director begins the important role of readying all camera shots from the technical director's copy of the updated master script. This will be the technical director's task during successive rehearsal(s) and eventually during videotaping. The technical director readies each successive camera shot by first calling the camera number (e.g., "ready camera 1") and the consecutive camera shot number (e.g., "shot number 3").

Readying all camera shots in advance from the control room (TD 9) The technical director readies as many camera shots from the technical director's copy of the master script ahead of time as possible without confusion (e.g., two or three). The technical director should spread consecutive pages of the master script across the top of the switcher for ready reference. Besides readying shots, the technical director must cut the readied shots when the director calls for them to be cut. With the three camera studio, the technical director merely needs three fingers on the three switcher buttons assigned to each of the three cameras.

Cutting from shot to shot as the director calls for each take during rehearsal(s) and videotaping (TD 10) The technical director cuts from camera to camera and shot to shot as the director calls each shot during rehearsal(s) and, finally, the videotaped take(s).

Calling a break for costuming and make-up (D 11) When the director feels all blocking, lines, camera shots, and sound coverage are adequate, the director calls a break for the actor(s) and actress(es) to get into costumes and make-up and a break for the production crew.

Meeting with the actor(s) and actress(es) for last minute instructions and changes in make-up design (MA 2) The make-up artist meets with the cast immediately into the studio break for costuming and make-up to convey any last minute instructions on make-up and any changes in make-up design.

Meeting with the make-up artist for make-up preparation and beginning make-up application (T 4) The cast meets with the make-up artist and begins make-up application when the director breaks from rehearsal(s) for make-up and costuming. Make-up application that is detailed and a slow and tedious process must be begun as soon as possible after cast call instead of waiting until the rehearsal break.

Meeting with the actor(s) and actress(es) to distribute costumes (CM 2) The costume master/mistress meets with the cast to distribute the costumes required for the script unit in production.

Meeting with the costume master/mistress for costume distribution (T 5) After make-up is complete, the cast meets with the costume master/mistress for the distribution of their costumes.

Beginning costuming (T 6) The cast begins costuming after applying make-up. Actor(s) and actress(es) usually costume after make-up; however, occasionally (e.g., if close fitting costumes must be drawn over the head and face), actor(s) and actress(es) will costume themselves and then apply make-up. Costumes need to be protected from make-up stains. Tissues stuck into the necks of costumes protect the costume materials from make-up stains until the last possible minute before videotaping.

Assisting actor(s) and actress(es) into costumes (CM 3) The costume master/mistress is responsible for the proper dressing of the cast and for the correct accessories for each costume. The costume master/mistress helps dress or supervises the cast as they dress.

Proceeding with make-up application (MA 3) The make-up artist helps or supervises the application of make-up by the cast. Most actor(s) and actress(es) apply their own make-up, but need a supervisory eye for amount and correctness of design.

Checking the make-up of each actor(s) or actress(es) before videotaping (MA 4) The make-up artist checks each member of the cast for final make-up design before each returns to the studio. The make-up artist is still responsible for the make-up designs approved during preproduction.

Watching takes for additional make-up needs, changes, or repairs (MA 5) The make-up artist monitors the videotaped takes for any make-up needs, changes, or powdering repairs the cast may require.

Being available yet out of the way in the studio during subsequent rehearsal(s) and take(s) (T 7) Once costuming and make-up are complete, the cast returns to the studio and awaits the director and final crew preparations. This usually means some waiting. A perennial problem in studio production is cast and crew getting in each other's way as final touches are made before a take. The cast should be readily available as needed by the director or the crew, but, at the same time, should stay out of the way of crew task completion.

Calling for a final rehearsal or a take (D 12) When the cast is finished with costuming and make-up and has returned to the studio, the director calls for a final rehearsal or a videotaped take. The director may videotape the rehearsal without alerting the studio, but a final rehearsal after costuming and make-up serve to check out the effects of the lighting on both. If the rehearsal or take is to be recorded, the director must ready the videotape recorder operator, await the response that the videotape recorder is ready, and then roll the videotape and take the slate for the unit. If the director has chosen not to be on the intercom network, the technical director relays the director's call to the videotape recorder operator.

Observing studio monitors for additional make-up needs, changes, or repairs (MA 6) The make-up artist observes studio monitors for any additional make-up needs, changes, or repairs to the actor(s) and actress(es) under the lights. Changes and repairs are made between rehearsal(s) and take(s).

Monitoring lighting of set(s), actor(s), and actress(es) over the control room monitors during take(s) (LD 6) The lighting director monitors the set lighting levels and cast lighting over the control room monitors during take(s).

Readying, rolling, and recording the videotape deck with the director's call (VTRO 5) The videotape recorder operator responds to the director's call to ready the recorder videotape deck. The videotape recorder operator then begins to roll the recorder deck and confirms to the director that the deck is recording.

Preparing and changing relevant character generator slate information for all videotaped takes of any script unit (PA 4) The production assistant prepares and readies the character generator for the slate copy or any other information needed by the director during videotaping.

Shooting camera for take(s) (CO 12) The camera operators respond to the director's call and shoot according to their shot lists for the required take(s). Camera operators follow their shot lists carefully and note any changes or additions to their camera placement, shot composition, or framing.

Monitoring the video levels of the cameras during videotaping (VEG 4) The video engineer constantly monitors the video levels of the light entering the lenses of the cameras during videotaping. The dynamic motion

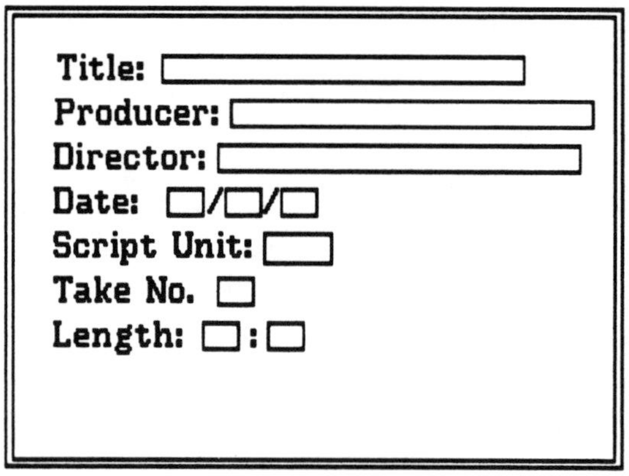

FIGURE 2–27
Character generator slate. The character generator slate is used to preface each videotaped take. The important information for the take is indicated on the slate.

of the cast and cameras during production frequently causes a burst of light into the lenses (e.g., an actor turning a highly reflective property may create a strong light reflection.) The video engineer must manually close the iris of the lens to control the burst of unexpected light.

Alerting the director to soft focused cameras during videotaping (VEG 5) The video engineer should alert the director and technical director that a camera may be soft focused. The video engineer keeps a closer watch on control room monitors of the studio cameras and can spot cameras with less than sharp focus. Alerting the director and technical director permits them to instruct the camera operator to correct the focus.

Recording any dialogue and mixing effects during take(s) while monitoring for any extraneous sounds during the take(s) (A 15) During the take(s), the audio director records dialogue and mixes any required sound effects. The audio director also monitors the studio for any extraneous and unwanted sounds during videotaping.

Covering the actor(s) and actress(es) for sound during take(s) (MBG/O 12) The microphone boom grip(s)/operator(s) cover all actor(s) and actress(es) for sound during studio take(s).

Following and keeping track of the master script for the director at the directing console in the control room during rehearsal(s) and take(s) (AD 7) The assistant director follows the master script for the director with a finger or pencil so the director can find the spot in the master script where the actor(s), actress(es), and dialogue are at any moment. The director depends on the assistant director to keep track of the dialogue.

Finishing a take and stopping videotape recording (D 13) At the end of a take, the director calls for a cut to the studio, a freeze, and a stop to the videotape recorder.

Stopping the videotape recorder at the director's call (VTRO 6) The videotape recorder operator stops down

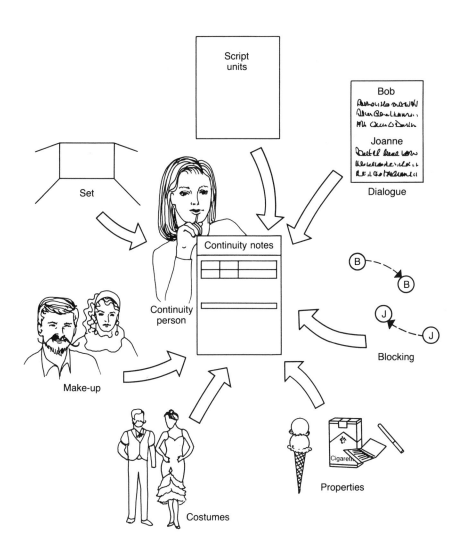

FIGURE 2–28
Continuity note taking. The continuity person takes notes on all elements of the production session.

the videotape recorder when the director calls for a stop to the videotaping.

Recording details of actor(s), actress(es), set(s), dialogue, properties, and costumes from unit to unit and take to take for reestablishing for a subsequent unit or take (CP 3) When the director calls for a cut, stops videotaping a take, and calls for a freeze, the continuity person enters the set while the cast freezes. The continuity person records all details that may affect subsequent units, takes, and, eventually, postproduction editing. Details of actor(s), actress(es), set(s), dialogue, properties, and costumes will have to be reestablished and matched for subsequent unit(s) and take(s) during production. The content of some edits will need to be matched during postproduction editing, and these continuity details will have to be cut together.

Releasing actor(s) and actress(es) after recording continuity details after a take or a wrap (CP 4) The continuity person releases actor(s) and actress(es) after recording continuity details. The cast is free to break only after the continuity person releases them.

Double checking the director's calls for rehearsal(s), retake(s), next unit for blocking and videotaping, ***studio wrap(s), or studio strike (AD 8)*** The assistant director assists the director by checking and relaying the director's commands throughout rehearsal(s) and tapings. This means that the assistant director knows what the director wants at any moment or intends to do at all times. This is especially true for announcing subsequent script units for blocking. The assistant director then informs others and is the resource person for that information.

Assisting the director in the control room and studio, keeping track of the production process, and updating the control room log of videotaped take(s) (AD 9) The assistant director assists the director in whatever needs the director may have in both the control room and the studio. The assistant director must keep track of the production process itself for the director. The assistant director keeps the control room log of videotaped take(s), which is a record of the source tape for postproduction editing needs. (See the videotape log form.)

Effecting what the director requires while the director is in the control room (F 5) The floor director continues to function as the director's head and hands in the studio. This means that anything the director needs for or from the studio should be handled by the floor director. Once the director leaves the studio for the control room, the director should not have to return except for a serious problem.

Informing self and the cast and crew of the director's intentions (F 6) The floor director is informed by the technical director over the intercom of the director's intentions at all times. The director's clear intentions can facilitate the actions of the cast and crew. The director may call for another rehearsal, a take, a retake, pickup shots, a wrap, the next unit for blocking, or a studio strike.

Assisting the director's judgment or call for a retake or wrap (TD 11) At the end of a take, the technical director assists the director's judgment of the quality of a take by sharing insight and alerting the director to any mistakes before the director calls for a retake or wrap.

Conferring with the producer on acceptable take(s) (D 14) The director should confer with the producer on the acceptability of any take before announcing the next call. Together the director and producer decide on which pickup shots from the set, actor(s)/actress(es), and props to use during postproduction editing. Pickup shots are used to cover mismatched video edits. The decision to wrap a unit should be a mutual decision of the producer and director.

Conferring with the director on the acceptability of take(s) (P 4) The producer confers with the director after each take and helps decide the acceptability of a take. The producer assists the director in deciding the content of pickup shots to be used in postproduction. Pickup shots are used to cover video shots that do not match in editing. The decision to wrap a script unit should be a mutual decision of the producer and director.

Replacing or replenishing properties consumed or used during rehearsal(s) or take(s) (PM 4) The properties master/mistress replaces or replenishes any properties that may have been consumed or used during a rehearsal or take. There are properties that will be eaten or smoked that will have to be replenished or relit before a subsequent rehearsal or take.

Replacing or altering properties to the changing needs of the teleplay during production (PM 5) During production, some properties will have to be changed or altered depending on the changing needs of the teleplay and the script. Some consumable properties, such as ice cream cones, will have to be adapted to a later stage in consumption to match a previous unit. Cigarettes, for example, will have to be lit at a half smoked stage for a later unit.

Watching rehearsal(s) and take(s) for the use, abuse, and needed repair to costumes (CM 4) The costume master/mistress watches rehearsal(s) and take(s) closely for the way actor(s) and actress(es) use and abuse the costumes and prepares for any needed repair of the costumes between rehearsal(s) and take(s).

Making adjustments and repairs to costumes (CM 5) The costume master/mistress monitors the use of the costumes and after rehearsal(s) and take(s) is ready to make adjustments and repairs.

Knowing the rehearsal and take schedule for the actor(s) and actress(es) and assisting in readying them for subsequent rehearsal(s) and take(s) (MA 7) The make-up artist knows the rehearsal and take schedule for the actor(s) and actress(es) so as to be better able to assist them in preparing for subsequent rehearsal(s) and take(s).

Checking with the continuity person between script units and takes (T 8) The cast always checks with the continuity person for reestablishing details of dialogue, costumes, make-up, and properties before beginning a subsequent unit or take. The actor(s) and actress(es) have to begin subsequent takes with the same details they ended a previous take to facilitate a postproduction edit required at that point.

Making frequent checks with the costume master/ mistress and make-up artist to review and repair costumes and make-up when necessary (T 9) Actor(s) and actress(es) check frequently with the costume master/ mistress and the make-up artist for costume and make-up review. If repair of either is needed, it should be made as soon as possible.

Removing old spike marks before blocking begins for a new script unit (F 7) The floor director begins to remove old spike marks from the studio floor when the director calls for a wrap for a previous script unit.

Overseeing the director's call for a take, a retake, next script unit, a studio wrap, or a studio strike (F 8) The floor director oversees the studio as the director makes a disposition call on a previous take. If the director wants another take, the floor director readies the cast and crew for the beginning of that take. If the director calls for pickup shots from the studio, the floor director must convey to the studio personnel exactly what the director wants videotaped. If the director calls for a wrap and the next script unit, the floor director prepares for that unit by announcing the director's intentions. If the director calls for a studio strike, the floor director oversees the striking of the studio and studio personnel.

Responding to the director's call for a retake, a wrap, next script unit, or a studio strike (TD 12) The technical director is ready for whatever the director's next call may be—a retake of the previous unit, pickup shots, a wrap for the previous unit, the next scheduled unit, or a studio strike.

Reestablishing the details of a previous take before beginning succeeding retake(s) or unit(s) (CP 5) If the director's call is to retake a unit, the continuity person instructs the actor(s) and actress(es) on the details of dialogue, costume, properties, and make-up as they were before the beginning of the previous take. These details ensure against unmatched edits in postproduction.

Announcing production intentions (D 15) The director needs to make a decision and announce the next production intention to the cast and crew. The director could choose to do another take, call a wrap for a particular script unit and the intention to block the next scheduled unit, or, at the end of the production session, call a studio strike. The director should make intentions clear to the assistant director and technical director who in turn alert those crew and cast subordinate to them or on their intercom networks.

Securing talent release signatures (P 5) The producer secures any required talent release signatures from the cast. If studio extras were employed for the production session, talent release signatures are secured from them before they depart the production facility.

Preparing to block the next script unit or striking the studio cameras (CO 13) The camera operators either prepare to begin the process of production again with the blocking of the next scheduled script unit or they strike the studio cameras by capping the lenses; locking pan, tilt, and pedestal; and recoiling the camera cables.

Preparing to light the next script unit or set or striking floor lights and turning off set lights at a strike call (LD 7) The lighting director prepares to light the next script unit set or begins to strike the floor lights and turns off the set lights at the director's strike call.

Preparing for the next script unit with a wrap call by the director or disassembling microphone(s), holder(s), and cable(s) and closing down the audio control board with a strike call by the director (A 16) The audio director prepares to continue sound coverage with the director's call for a wrap on one script unit and the announcement of another. The studio microphone(s), holder(s), and cables will be struck if the director calls for a strike.

Preparing for the next script unit with a wrap call by the director or disassembling the microphone(s), boom(s), and cables with a strike call by the director (MBG/O 13) The microphone boom grip(s)/operator(s) prepare to continue sound coverage with a wrap call for one script unit and the announcement of the next unit. If the director calls for a studio strike, the grip(s)/operator(s) begin disassembling the microphone(s), boom(s), and cables.

Overseeing striking the control room and securing the director's master script and videotape log form (AD 10) The assistant director conveys the director's call to strike to the crew, oversees the strike of the control room, and secures the director's master script and videotape log until the next production session.

Overseeing the complete strike of the production studio (F 9) The floor director oversees the complete strike of the studio. This means monitoring the storage of cameras, breakdown of studio hardware, shutting down the set, sorting properties, and, in general, cleaning up after the crew and cast.

Capping the cameras with the director's call for a studio wrap and strike (VEG 6) The video engineer responds to the director's call for a wrap and strike by electronically capping the studio cameras from the camera control area.

Shutting down the switcher and securing the master script copy (TD 13) The technical director shuts the switcher down and secures the technical director's copy of the master script for the next production session.

Rewinding source videotape stock and accurately labeling the content of the videotape on the videotape and its storage case (VTRO 7) The videotape recorder operator, after shutting the record deck down, rewinds the source tape and accurately labels the videotape and its storage case with the content of the session's tapings.

Shutting down the character generator and securing memory record of its content (PA 5) The production assistant shuts the character generator down with the director's call for a wrap. The production assistant then secures the memory list of recorded character generator content for the next production session.

"Break 10 . . ."
"Next unit blocking . . ."
"Keep going . . . End board"
"Take . . ."
"Posting . . ."
"Retake . . ."
"Pickup shots . . ."
Director
"That's a wrap . . ."
"Strike the studio . . ."

FIGURE 2–29
Director's production session intentions. The director is expected to make successive production intentions known to the cast and crew as soon as decisions are made.

Reordering continuity forms (CP 6) The continuity person reorders all of the continuity forms recorded for the production session into the order of the takes. Forms needed to reestablish the next session's production units should be secured.

Collecting, cleaning, and storing properties after take(s) and studio strike call (PM 6) The properties master/mistress collects properties from the cast and the set for cleaning if needed and storing until the next production session.

Retrieving, repairing, and cleaning costumes used during production (CM 6) The costume master/mistress retrieves costumes from the actor(s) and actress(es) after the studio wrap is called. Some costumes will have to be repaired and some cleaned before the next production session. Make-up residue is always a problem for costumes, so they require regular cleaning.

Assisting actor(s) and actress(es) in removing make-up and cleaning up after a strike (MA 8) The make-up artist assists the actor(s) and actress(es) in removing their make-up. Detailed make-up (e.g., that for aging or facial hair) requires the assistance of the make-up artist, especially if elements of the make-up are to be reused. Of all areas of production, make-up requires the most effort to clean up, not only the cast but also the make-up facilities.

Removing costumes and make-up and cleaning up (T 10) Actor(s) and actress(es) remove costumes and make-up after the director's call for a strike. The costume master/mistress and make-up artist assist the cast. The cast should report costume repair and cleaning needs. Cleaning up costume and make-up facilities is a shared responsibility among the cast and their staff.

THE POSTPRODUCTION PROCESS

The postproduction process is an important step in the production of television studio drama. The production of studio drama was accomplished by breaking a teleplay into producible script units. These units, videotaped during the production process—often out of teleplay order—now must be edited together into the complete teleplay.

At the beginning of postproduction, the producer and director have the source videotapes from the videotape recorder operator, labeled and ordered by studio production session. These source tapes and adequate master videotape stock should also have been striped by the videotape recorder operator with SMPTE time code. The director and producer also have the videotape log that was kept by the assistant director during studio production on which was recorded all of the videotape takes of all script units. Finally, the director has the master script, which was updated during production and reordered into the order of the original teleplay. The audio director supplies any remaining music or sound effects not already wedded to the source videotapes. Postproduction can begin.

• Personnel

Personnel required for the postproduction of the studio drama will vary by the custom of a facility and the skills of the production personnel. A producer continues in the overseeing role as defined by the producer's role in production. The director may choose to do the postproduction editing. On the one hand, the best person to do the postproduction editing is the director, who designed the production and directed the studio stages of the production. On the other hand, it would not be uncommon for a technical director/editor to serve as a videotape editor. If a technical director/editor will do the editing, the producer and director work with the technical director/editor during the editing sessions.

Producer (P) The producer from the preproduction and production stages continues in that role in postproduction. The role continues as a supervisory and organizing role. The producer attends to scheduling and creative consulting before and during editing.

Director (D) The director continues into postproduction as the catalyst in seeing the teleplay through to completion. In some operations, the director will do the editing. In others, a technical director/editor will do the technical editing under the guidance and creative direction of the director.

Technical director/editor (TD/E) If the director of the production does not do the editing of the source tapes, a technical director/editor will be called upon to do it. This title and role is unrelated to the control room technical director of the production process. This role is generally a button-pushing role under the guidance and creative direction of the director with assistance from the producer.

Audio director (A) The audio director plays a role in postproduction insofar as the sound track of the final edited master tape will need postproduction work. Primarily, the music and sound effects track will have to be laid and mixed during postproduction. While a technical director/editor has the skills to do this editing, the audio director must supply the audio sources. When the video portion of the master tape is completely edited, the audio director may be required (when available technology and required degree of skill permit) to separate the dialogue and sound effects tracks from the master, sweeten the master, and mix the remaining music and sound effects into the master tape.

Assistant director (AD) The assistant director may be required to perform a continuing assistance role to the director during postproduction. The assistant director was very close to the director's master script during production and also logged all videotape takes during production. This experience can be an invaluable resource to the producer and director (and technical director/editor) during postproduction.

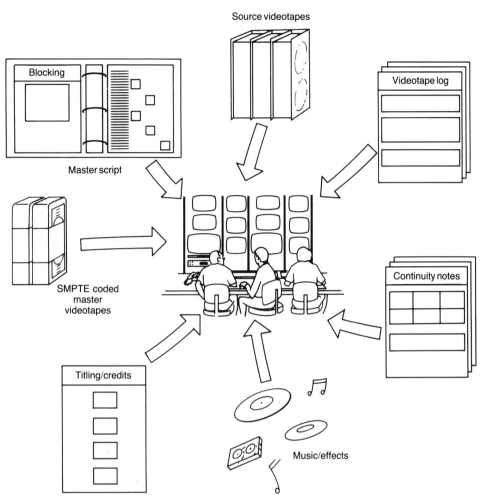

FIGURE 2–30
Postproduction preparation. Many elements have to come together before postproduction editing can begin.

Source videotapes

Blocking

Master script

Videotape log

SMPTE coded master videotapes

Continuity notes

Titling/credits

Music/effects

Videotape recorder operator (VTRO) The videotape recorder operator plays a limited role in postproduction. The videotape recorder operator possesses the source videotapes, has labeled them after production, and has striped each one with SMPTE time code. The master videotape stock to be used in postproduction should also be striped with time code.

• **Postproduction Stages**

Arranging the postproduction editing schedule and facilities (P 1) The producer begins postproduction by arranging for the editing facilities and scheduling editing time. The schedule and available editing facilities will have to account for the availability of a videotape editor. Either the director of the production or a technical director/editor will do the editing.

Supervising postproduction preparation (P 2) The producer continues postproduction by supervising preparation for editing. This means that resources including source and master videotapes have to be secured, the master script reordered, and the videotape take log secured.

Supervising postproduction preparation (D 1) The director of the production shares in supervising postpro-

duction preparation with the producer. Many of the details of the production required for postproduction were under the direction of the director (e.g., the master script). (See the postproduction cue sheet form.)

Reordering and editing the master script (D 2) The director prepares the master script for postproduction editing. The master script is the primary resource for postproduction editing (after the source videotapes). The master script will probably require some reordering back into the original order of the teleplay. During blocking and videotaping, the teleplay was broken into units and produced in the studio in a different production order from the teleplay script order. The master script will have to be reordered for postproduction. There may be other editing decisions to be made before postproduction editing in terms of production problems or dramatic interpretations that occurred during preproduction and production. Included among these decisions is the use of stock shots and pickup shots videotaped during production.

Providing music and sound effects tracks (A 1) The audio director is responsible for providing any music or sound effects tracks for postproduction. Some music and effects are already a part of the source tapes. Other music and effects must be made across edits in postproduction. (See the effects/music cue sheet form.)

Turning the videotape take log forms over to the producer (AD 1)   The assistant director turns the videotape take log forms over to the producer for postproduction reference. The videotape take logs are a detailed resource of the order and disposition of each take on the source tapes.

Striping the source and master videotapes (VTRO 1)
The videotape recorder operator stripes all videotape stock with SMPTE time code in preparation for editing. The time code permits an accurate determination of edit points as videotape editing is performed from the source tapes to the master tape during postproduction.

Turning the source videotapes and the master videotape stock over to the producer (VTRO 2)   The videotape recorder operator turns all videotapes over to the producer. The videotapes include the source tapes recorded during production and the videotape stock necessary for the master videotape.

Editing a rough cut edition (D 3)   The director begins a rough cut edit of the teleplay. A rough cut serves a number of purposes. It permits the director to sense the flow and pacing of the dialogue and plot across edits. As individual units of the teleplay are produced it is difficult to impossible to capture a sense of the whole plot.

A rough edit reveals where editing can control pacing: More rapid editing can pick up the pacing, and fewer edits will slow down the pacing. A rough edition will also reveal where tighter edits will serve the plot and dialogue. Shaving frames off at edit points creates a sharper perception of multiple camera production.

The length of the teleplay may be important. If the teleplay has to hit broadcast time slots, a rough cut allows an evaluation of length requirements and areas that can be lengthened or shortened.

Above all, a rough edit gives the director a sense of the whole, something not yet experienced. The rough edit becomes a point of reference for the final edit.

Observing the editing session with the director and assisting in edit decision making (P 3)   The producer keeps the editing schedule with the director and becomes a partner with the director in editing decisions. After the director, the producer is closest to the creative design of the teleplay and has much to share in editing decisions during postproduction. The final decision, however, should remain with the director.

Assisting the director during postproduction (AD 2)   The assistant director can be of service to the director during postproduction. A major task during editing is the need to search out units from the source videotapes for edit or the need to review other takes. The assistant director can help the director (or technical director/editor) and save editing time by searching out source tapes for takes while editing continues.

Editing from the master script and the videotape log forms under the supervision of the producer and director (TD/E 1)   If the technical director/editor does the editing in postproduction, the technical director/editor edits from the master script and the videotape log forms. The first editing is a rough edit. The technical director/editor works under the supervision of the producer and director.

Reviewing the rough edit (D 4)   The director reviews the rough edit of the teleplay with the producer as a step toward making decisions affecting the final edit. Edit pacing has to be evaluated, music and sound effects planned, additional titling and credits created, and broad-

FIGURE 2–31
Stages in postproduction. Postproduction stages include rough editing, final editing, and audio sweetening.

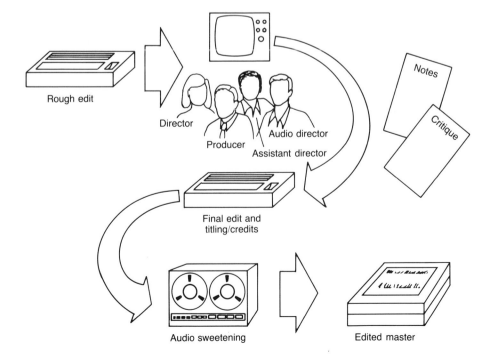

Rough edit

Director

Producer

Audio director

Assistant director

Notes

Critique

Final edit and titling/credits

Audio sweetening

Edited master

cast time block created. This step is a time for copious note taking as a prelude to undertaking a final edit.

Reviewing the rough edit (P 4) The producer reviews the rough edit with the director as the opportunity to recommend changes to the final edit, placement of edits, cut-aways, music, and sound effects. This is the first opportunity to experience the teleplay as a whole since reading the script.

Assisting the director and technical director/editor with audio problems and additions to the final edit (A 2) The audio director should plan to assist the director or technical director/editor with any decisions that involve the audio portion of the teleplay. Given the expertise of the audio director, some questions about the sound track are more easily solved with the input of the audio director.

The audio director reviews the rough edit with the director.

Making the final edit (D 5) The director begins the final edit after reviewing the rough edit with the producer and audio director. The final edit should embody the ideas and improvements suggested by the producer and audio director.

The director adds music and sound effects to the second channel and mixes the audio channels to a single channel. The audio director may be required to sweeten the audio track with more polished audio postproduction.

The director also adds any titling or credits that may not have been matted during videotaping.

Making the final edit, adding music and sound effects, and mixing the audio channels (TD/E 2) If a technical director/editor is making the final edit, the technical director/editor adds music and sound effects to the second audio channel and mixes the audio channels to a single channel. The audio director may be required to sweeten the audio track with more polished audio postproduction.

Sweetening the sound track of the final edit (A 3) The audio director may be required to sweeten the entire sound track of the final edit as a remaining postproduction responsibility. This will entail separating the time coded sound track from the time coded video track. Once this is done, the sound track can be worked with (e.g., the audio reproduction of the studio sound can be cleaned up or sound effects and music added). The sound track is then rewedded to the video track with accurate time coding.

Labeling the edited master videotape and removing the record button (D 6) When video editing is complete and audio sweetening done, the director labels the final edited master videotape and removes the record button to prevent accidental erasing of the video.

Labeling the edited master videotape and removing the record button (TD/E 3) If the technical director/editor did the editing, the technical director/editor labels the final edited master videotape and removes the record button to prevent accidental erasing of the video.

Chapter 3

Studio Production Organizing Forms

There are a diversity and breadth of roles and task responsibilities in television studio drama producing and production unknown in other television genres. This very diversity and breadth warrant some technique within which role and task responsibility of personnel can be successfully met. It is frequently said in television production that the success of a production is a function of the extent and degree of preproduction.

Organizing forms are presented in this chapter for almost all information gathering and task preparation stages required for studio drama preproduction and production. The majority of the forms serve to design and order drama producing tasks. The remaining forms help to prepare for the actual studio production of the teleplay. Included are forms that were intended to assist in the design of drama production and in the development of the cast from audition to characterization. As with many resources in a creative and technological medium, not all forms will be equally functional. These forms should be used when they facilitate the tasks for which they were created. They should not become ends in themselves. Forms that serve the production should be used and perhaps adapted to the requirements of a particular teleplay or studio production facility. Still others may be needed and may serve a one time only function. Others may not be required and should be overlooked.

The forms presented here are designed to be used by producing and production personnel for particular role accomplishment. Those roles are listed in the upper portions of the respective forms. These forms can be removed or photocopied. Note that each form's structure and function are detailed in Chapter 4.

PRODUCTION BUDGET
MULTIPLE CAMERA VIDEO DRAMA PRODUCTION

Producer: Teleplay Title:
Director: Author:
Date: /   /   Length: :

No. Preproduction Days: [   ] Hours: [   ]   First Production Date: /   /
No. Studio Shoot Days: [   ] Hours: [   ]   Completion Date: /   /
No. Postproduction Days: [   ] Hours: [   ]

SUMMARY OF PRODUCTION COSTS	ESTIMATED	ACTUAL
1. Story Rights, Other Rights and Clearances		
2. Teleplay Script		
3. Producer and Staff		
4. Director and Staff		
5. Talent		
6. Benefits		
7. Production Facility		
8. Production Staff		
9. Camera, Videotape Recording, Video Engineering		
10. Set Design, Construction, Decoration		
11. Set Lighting		
12. Properties		
13. Costuming		
14. Make-up and Hairstyling		
15. Audio Production		
16. Videotape Stock		
17. Postproduction Editing		
18. Music		
19. Photography		
20. Subtotal		
21. Contingency		
Grand Total		

COMMENTS

STORY RIGHTS, OTHER RIGHTS AND CLEARANCES	ESTIMATED	ACTUAL
1. Story Rights Purchase		
2.		
3. Other Rights		
4. Music Clearances		
5.		
Subtotal		

TELEPLAY SCRIPT	ESTIMATED	ACTUAL
6. Writer Salary No. ( ) @ ( )		
7.		
8. Story Editor No. ( ) @ ( )		
9.		
10. Secretary: No. ( ) @ ( )		
11. Searches (authenticity, libel avoidance, etc.)		
12. Office Supplies		
13. Travel Expenses		
14. Photocopying		
15. Telephone		
16. Postage		
17. Miscellaneous		
Subtotal		

PRODUCER AND STAFF								
CREW	ESTIMATED				ACTUAL			
	Days	Rate	O/T Hrs	Total	Days	Rate	O/T Hrs	Total
18. Producer								
19. Preproduction								
20. Production								
21. Postproduction								
22.								
23. Secretary: No. ( )								
24.								
25. Producer Supplies								
26. Photocopying								
27. Telephone								
28. Postage								
29.								
30. Travel Expenses								
31. Per Diem								
32. Miscellaneous								
33.								
Subtotal								

DIRECTOR AND STAFF

CREW	ESTIMATED				ACTUAL			
	Days	Rate	O/T Hrs	Total	Days	Rate	O/T Hrs	Total
34. Director								
35. Preproduction								
36. Production								
37. Postproduction								
38. Assistant Director								
39. Preproduction								
40. Production								
41. Postproduction								
42.								
43. Secretary: No. ( )								
44. Director's Supplies								
45. Photocopying								
46. Telephone								
47. Postage								
48. Expenses								
49. Miscellaneous								
50.								
Subtotal								

TALENT

DESCRIPTION	ESTIMATED				ACTUAL			
	Days	Rate	O/T Hrs	Total	Days	Rate	O/T Hrs	Total
51. Cast								
52. CHARACTER ｜ ACTOR								
53.								
54.								
55.								
56.								
57.								
58.								
59.								
60.								
61.								
62.								
63.								
64.								
65.								
66.								
67.								
68.								
69.								
70.								
71. Stunts								
72. Casting Service Fee								
73. Extra Talent								
74.								
Subtotal								

BENEFITS

DESCRIPTION	TOTAL
75. Insurance Coverage	
76. Health Plan	
77. Welfare Plan	
78. Taxes (FICA, etc.)	
79. Equity/Guild/Union Costs	
80.	
Subtotal	

PRODUCTION FACILITY

SPACE	ESTIMATED				ACTUAL			
	Days	Rate	O/T Hrs	Total	Days	Rate	O/T Hrs	Total
81. Production Studio								
82. Studio Camera No.( )								
83. Teleprompter								
84. Floor Monitor No.( )								
85.								
86. Control Room								
87. Production Switcher								
88. Digital Video Effects								
89. Still Store								
90.								
91. Intercom Network								
92.								
93. Character Generator								
94.								
95. Audio Control Board								
96. Cartridge Playback								
97. Cassette Playback								
98. Reel-to-Reel Playback								
99. Turntable								
100. Compact Disc Player								
101. Studio Foldback								
102.								
103. Master Control								
104. Record Deck No.( )								
105. Playback Deck No.( )								
106. Other								
107.								
108. Dressing Rooms								
109. Costume Storage Room								
110. Make-up Facility								
111. Green Room								
112. Other								
Subtotal								

PRODUCTION STAFF

STAFF	ESTIMATED				ACTUAL			
	Days	Rate	O/T Hrs	Total	Days	Rate	O/T Hrs	Total
113. Technical Director								
114. Preproduction								
115. Production								
116. Postproduction								
117. Floor Director								
118. Preproduction								
119. Production								
120. Postproduction								
121. Production Assistant								
122. Preproduction								
123. Production								
124. Postproduction								
125. Continuity Person								
126. Preproduction								
127. Production								
128. Postproduction								
129.								
130. Other Crew								
131.								
132. Miscellaneous								
133.								
134.								
Subtotal								

CAMERAS, VIDEOTAPE RECORDING, AND VIDEO ENGINEERING

DESCRIPTION	ESTIMATED				ACTUAL			
	Days	Rate	O/T Hrs	Total	Days	Rate	O/T Hrs	Total
135. Camera Operator No. ( )								
136. Preproduction								
137. Production								
138. Postproduction								
139. VTR Operator								
140. Preproduction								
141. Production								
142. Postproduction								
143. Video Engineer								
144. Preproduction								
145. Production								
146. Postproduction								
147. Equipment Rental								
148. Equipment Purchases								
149. Maintenance/Repair								
150. Miscellaneous								
151.								
Subtotal								

SET DESIGN, CONSTRUCTION, AND DECORATION

DESCRIPTION	ESTIMATED				ACTUAL			
	Days	Rate	O/T Hrs	Total	Days	Rate	O/T Hrs	Total
152. Set Designer								
153.								
154. Construction Labor								
155.								
156. Construction Materials								
157.								
158. Miscellaneous								
159.								
160. Set Dressing Labor								
161.								
162. Set Dressing Props								
163. Cleaning/Loss/Damage								
164. Purchases								
165. Rentals								
166.								
167. Miscellaneous								
168.								
	Subtotal							

PROPERTIES

DESCRIPTION	ESTIMATED				ACTUAL			
	Days	Rate	O/T Hrs	Total	Days	Rate	O/T Hrs	Total
169. Master/Mistress								
170.								
171. Property Labor								
172. Vehicle								
173. Transportation								
174. Storage								
175.								
176. Animal(s) No. ( )								
177. Animal Handler								
178. Feed/Stabling								
179.								
180. Action Properties								
181. Rented								
182. Purchased								
183. Cleaning/Loss/Damage								
184.								
185. Hand Properties								
186. Rented								
187. Purchased								
188. Storage								
189. Food								
190.								
191. Miscellaneous								
	Subtotal							

SET LIGHTING

DESCRIPTION	ESTIMATED				ACTUAL			
	Days	Rate	O/T Hrs	Total	Days	Rate	O/T Hrs	Total
192. Lighting Director								
193. Preproduction								
194. Production								
195. Postproduction								
196.								
197. Expendables								
198. (gels, etc.)								
199. Lighting Equipment								
200. Rental								
201. Purchases								
202. Light Instruments								
203. 1 kw Spot No.( )								
204. 2 kw Spot No.( )								
205. kw Spot No.( )								
206. 1 kw Scoop No.( )								
207. 1 1/2 kw Scoop No.( )								
208. 2 kw Scoop No.( )								
209. kw Scoop No.( )								
210. Ellipsoidal Spotlight								
211. Follow Spot								
212. Broad/Softlight No.( )								
213. Strip/Cyc Light No.( )								
214. Miscellaneous								
215.								
	Subtotal							

AUDIO PRODUCTION

DESCRIPTION	ESTIMATED				ACTUAL			
	Days	Rate	O/T Hrs	Total	Days	Rate	O/T Hrs	Total
216. Audio Director								
217. Preproduction								
218. Production								
219. Postproduction								
220. Mike Grip No.( )								
221. Preproduction								
222. Production								
223. Postproduction								
224. Boom Operator No.( )								
225. Preproduction								
226. Production								
227. Postproduction								
228. Miscellaneous Labor								
229. Audio Boom No.( )								
230. Fishpole No.( )								
231. Audio Equipment								
232. Rental								
233. Purchases								

AUDIO PRODUCTION (Continued)

DESCRIPTION	ESTIMATED				ACTUAL			
	Days	Rate	O/T Hrs	Total	Days	Rate	O/T Hrs	Total
234. Microphones								
235. Shotgun No.( )								
236. Lavaliere No.( )								
237. Handheld No.( )								
238. Boom No.( )								
239. Wireless No.( )								
240. Desk No.( )								
241. Stand No.( )								
242.								
243. Miscellaneous								
	Subtotal							

COSTUMING

DESCRIPTION	ESTIMATED				ACTUAL			
	Days	Rate	O/T Hrs	Total	Days	Rate	O/T Hrs	Total
244. Master/Mistress								
245. Preproduction								
246. Production								
247. Postproduction								
248. Costume Labor								
249. Costumes								
250. Rental								
251. Purchases								
252.								
253. Loss/Damage								
254. Cleaning								
255. Costume Building								
256.								
257. Miscellaneous								
258.								
	Subtotal							

MAKE-UP AND HAIRSTYLING

DESCRIPTION	ESTIMATED				ACTUAL			
	Days	Rate	O/T Hrs	Total	Days	Rate	O/T Hrs	Total
259. Make-up Artist								
260.								
261. Make-up Supplies								
262.								
263. Hairstylist								
264. Hairstyling Supplies								
265. Wig								
266. Rental								
267. Purchases								
268. Miscellaneous								
269.								
	Subtotal							

VIDEOTAPE STOCK

DESCRIPTION	ESTIMATED				ACTUAL			
	Days	Rate	O/T Hrs	Total	Days	Rate	O/T Hrs	Total
270. Videotape:								
1" (:) x ($)								
3/4" (:20) x ($)								
3/4" (:30) x ($)								
3/4" (:60) x ($)								
271.								
	Subtotal							

POSTPRODUCTION EDITING

DESCRIPTION	ESTIMATED				ACTUAL			
	Days	Rate	O/T Hrs	Total	Days	Rate	O/T Hrs	Total
272. Technical Director								
273.								
274. Editing Suite Rental								
275. Playback/Search Deck								
276. Master Tape Stock								
277. SMPTE Coding								
278. Stock Video Footage								
279. Audio Sweetening Room								
280. Sound Effects								
281. Dubs ( ) x ($)per								
282. Miscellaneous								
283.								
	Subtotal							

MUSIC

DESCRIPTION	ESTIMATED				ACTUAL			
	Days	Rate	O/T Hrs	Total	Days	Rate	O/T Hrs	Total
284. Music Purchases								
285. Music Royalties								
286.								
287. Song Writer								
288. Arranger								
289. Composer								
290. Conductor								
291. Vocalist No.( )								
292.								
293. Recording Facility								
294. Recording Crew								
295. Audio Recording Tape								
296.								
297. Miscellaneous								
298.								
	Subtotal							

PHOTOGRAPHY

DESCRIPTION	ESTIMATED				ACTUAL			
	Days	Rate	O/T Hrs	Total	Days	Rate	O/T Hrs	Total
299. Photographer								
300. Film Stock								
301. Film Processing								
302. Printing								
303.								
304.								
	Subtotal							

COMMENTS

FACILITY REQUEST

MULTIPLE CAMERA VIDEO DRAMA PRODUCTION

Production Facility:

Producer: **Teleplay Title:**
Director: **Teleplay Length:**   :
Date: /   /   **Production Dates:**   /   /   to   /   /   **Hours:**   :   to   :

Facility Space:
☐_____
☐ Studio   ☐ Audio Control Room   ☐ Set Construction Workshop
☐ Control Room   ☐ Make-Up Room   ☐ Costume Storage
☐ Master Control   ☐ Dressing Room   ☐ Property Storage

Studio Requirements:
☐ No. Studio Cameras   ☐ Teleprompter   ☐ No. Floor Monitors
   Light Instruments   ☐_____   ☐ Audio Boom Platform
   ☐ __kw spots   ☐ __kw scoops   ☐ ellipsoidal spotlight   ☐ follow spot   ☐ broad/softlight
   ☐ __kw spots   ☐ __kw scoops   ☐ strip/cyc lights   ☐ other_____

Control Room Requirements:
☐_____
☐ Production Switcher   ☐ Character Generator   ☐ Intercom Network: no. stations___
☐ DVE   ☐ Still Store/E-men   ☐ Light Control Board

Master Control Area Requirements:
☐ Videotape Decks   ☐ Telecine   ☐_____
     ☐ record machines   ☐ 35mm slides
     ☐ playback machines   ☐ 16mm

Audio Control Room Requirements:
☐ Audio Control Board   ☐ Studio Foldback   ☐_____
☐ Playback Input Units:
     ☐ audio cartridge   ☐ audio cassette   ☐ reel-to-reel   ☐ turntable   ☐ compact disc
☐ Microphones:
     ☐ lavaliere   ☐ handheld   ☐ boom   ☐ wireless   ☐ desk   ☐ stand

Preproduction Requirements:
   Set Construction/Decoration: Date: /   /   Hours: :   to   :
                                    Date: /   /   Hours: :   to   :
                                    Date: /   /   Hours: :   to   :
   Set Lighting: Date: /   /   Hours: :   to   :
                  Date: /   /   Hours: :   to   :
                  Date: /   /   Hours: :   to   :

Studio Production Personnel:

Production Role	Personnel	Facility Crew	Program Crew
Director		☐	☐
Assistant Director		☐	☐
		☐	☐
Production Assistant		☐	☐
		☐	☐
Technical Director		☐	☐
Lighting Director		☐	☐
		☐	☐
Audio Director		☐	☐
Microphone Boom Operator		☐	☐
Microphone Grip		☐	☐
		☐	☐
		☐	☐
Floor Director		☐	☐
		☐	☐
Teleprompter Operator		☐	☐
		☐	☐
Camera Operator 1		☐	☐
Camera Operator 2		☐	☐
Camera Operator 3		☐	☐
Camera Operator 4		☐	☐
		☐	☐
Telecine Operator		☐	☐
		☐	☐
Videotape Recorder Operator		☐	☐
		☐	☐
Video Engineer		☐	☐
		☐	☐
		☐	☐
		☐	☐

TALENT AUDITION
MULTIPLE CAMERA VIDEO DRAMA PRODUCTION

Teleplay Title:	Author:

Actor/Actress:	Telephone:
Address:	Home ( ) -
City:	Work ( ) -
State: Zip Code:	Birthdate: / /

Availability	Unavailability
Dates: Hours:	Dates: Hours:

Personal Data: Height   Weight   Ethnicity:
  Hat Size: Shoe Size: Waist Measurement: Chest Measurement:
  In Seam: Shirt/Blouse Size: Suit/Dress Size:

Agent:
Union Affiliation:

Please explain your interest in this production:

Please list the roles/parts for which you feel qualified/interested:

Are you willing to perform as an extra for this production:

Are there other production tasks/roles in which you might be interested (e.g., costumes, make-up, properties, etc.):

PLEASE ATTACH YOUR RESUME AND PHOTO TO THIS FORM, OR LIST PREVIOUS EXPERIENCE
AND/OR DRAMA TRAINING ON THE REVERSE SIDE OF THIS FORM.

CHARACTERIZATION
MULTIPLE CAMERA VIDEO DRAMA PRODUCTION

Director: Teleplay Title:
                                  Author:
Actor/Actress: Date: /   /

Character:

List descriptive adjectives:

Order of importance	Order revealed to the audience
1.	1.
2.	2.
3.	3.
4.	4.
5.	5.
6.	6.
7.	7.
8.	8.

List actions performed by the character/done to the character:

Action	Significance
1.	1.
2.	2.
3.	3.
4.	4.
5.	5.

What are the major motivations of the character by script unit(s):
Unit(s)   :
Unit(s)   :
Unit(s)   :

What is this character's relationship to all other characters:

Character	Relationship
1.	
2.	
3.	
4.	
5.	
6.	

Characterization (Continued)

Label and elaborate on the function the character plays in the teleplay:

Describe the audience's emotional reaction to the character at important moments during the teleplay:

The Moment   Audience's Emotional Reaction

Describe the change in the character from the beginning to the end of the teleplay:

How can the change(s) be conveyed to the audience:

Write a birth to death biography of the character:

SCRIPT BREAKDOWN

MULTIPLE CAMERA VIDEO DRAMA PRODUCTION

Producer:	Teleplay Title:
	Author:
Director:	Length: :
	Script Length: pages
	Page of
	Date: / /

SCRIPT UNIT	SCRIPT PAGES/ LINES	INT/ EXT	TIME	SET	PROPERTIES	CAST	SHOOT- ING ORDER

SCRIPT UNIT	SCRIPT PAGES/ LINES	INT/ EXT	TIME	SET	PROPERTIES	CAST	SHOOT- ING ORDER

SET DESIGN

MULTIPLE CAMERA VIDEO DRAMA PRODUCTION

Set Designer:
Producer: Director: Teleplay Title:
   Approval ☐   Approval ☐   Author:
                                                  Date: /   /   Page   of

SET: SCRIPT UNIT(S):

Bird's Eye View   21 Units x 34 Units

Front View   12 Units x 34 Units

COSTUME DESIGN

MULTIPLE CAMERA VIDEO DRAMA PRODUCTION

Costume Master/Mistress:	Teleplay Title: Author:
Producer: Director: Approval ☐ Approval ☐	Date: / / Page of

Character:
  Script Unit(s):   Costume No.:

COSTUME DESCRIPTION   COSTUME SKETCH

Historical Period:
Costume Type:
Fabric:
Color Scheme:
Trim:
Costume Elements List:
       1.
       2.
       3.
       4.
       5.
       6.
Costume Accessories:
       1.
       2.
       3.
       4.
       5.

NOTES

MAKE-UP DESIGN

MULTIPLE CAMERA VIDEO DRAMA PRODUCTION

Make-Up Artist:	Teleplay Title:
	Author:
Producer: Director:	Date: / /
Approval ☐ Approval ☐	Page of

Character:	Actor/Actress:	☐ Wig:
Age: Complexion:		☐ Moustache:
Sex: M/F Type:		☐ Beard:

MAKE-UP ELEMENTS	TYPE/CODE		NOTES
Make-up base			
Highlighting			
Shadow			
Rouge			
Eye shadow			
Eye liner			
Eyebrow pencil			
Eyelashes			
Lip rouge			
Lip liner			
Highlighting liner			
Powder			
Liquid latex			
Crepe Hair			

Required make-up changes:
Script unit(s):   Script unit(s):
Make-up changes: Make-up changes:

Special requirements:

PROPERTIES BREAKDOWN

MULTIPLE CAMERA VIDEO DRAMA PRODUCTION

Properties Master/Mistress:

Producer: Director:
   Approval ☐   Approval ☐

Teleplay Title:
Author:
Script Length: pages
Script Units: units
Date : / /   Page   of

SCRIPT UNIT	SCRIPT PAGES/ LINES	CHARACTER(S)	PROPERTIES	NOTES

Insurance Required:
  Property: Type of Insurance:
  Property: Type of Insurance:

Properties Breakdown (Continued)				Page of
SCRIPT UNIT	SCRIPT PAGES/ LINES	CHARACTER(S)	PROPERTIES	NOTES

BLOCKING PLOT

MULTIPLE CAMERA VIDEO DRAMA PRODUCTION

Director:

Date: / /

Set:

Teleplay Title:

Author:

Script Page [   ]   Script Unit [   ]

Lighting
- ☐ Interior
- ☐ Exterior
- ☐ Day
- ☐ Night

Sound
- ☐ Synchronous
- ☐ Silent
- ☐

Bird's Eye Floor Plan

Cameras /Properties/Blocking

Description of Take/Unit   Actor(s)/Actress(es)/Cameras/Movement/Properties

In-Cue   Dialogue/Action   Out-Cue   Dialogue/Action

Comments:

Blocking Plot (Continued)		Page	of

Set:		Script Page ☐	Script Unit ☐

Lighting
☐ Interior
☐ Exterior
☐ Day
☐ Night

Sound
☐ Synchronous
☐ Silent
☐

Bird's Eye Floor Plan　　　　　Cameras /Properties/Blocking

Description of Take/Unit	Actor(s)/Actress(es)/Cameras/Movement/Properties

In-Cue	Dialogue/Action	Out-Cue	Dialogue/Action

Comments:

LIGHTING PLOT

MULTIPLE CAMERA VIDEO DRAMA PRODUCTION

Lighting Director:
Producer: Director: Teleplay Title:
    Approval ☐   Approval ☐   Author:
Set: Script Page ☐   Script Unit ☐
                                                          Date: /   /   Page   of

Lighting
☐ Interior
☐ Exterior
☐ Day
☐ Night

Lighting
    Change
☐ Yes
☐ No

Bird's Eye Floor Plan   Cameras /Properties/Blocking

Description of Take/Unit   Actor(s)/Cameras/Movement/Properties

Lighting Instruments   Filters   Property Lights
    Key Lights: Spun Glass: Lamps:
    Fill Lights: Gels: Ceiling:
    Soft Lights: Scrims: Other:
Lighting Accessories   Windows
    Barn Doors: Daylight:
    Flags: Night time:
    Gobo: Dusk:
                                  Change:

In-Cue   Dialogue/Action   Out-Cue   Dialogue/Action
                     Lighting Change Cue   Lighting Change Cue

Script Page [　　]　　　Script Unit [　　]

Lighting
☐ Interior
☐ Exterior
☐ Day
☐ Night

Lighting
　　　Change
☐ Yes
☐ No

Bird's Eye Floor Plan　　　　　　Cameras /Properties/Blocking

Description of Take/Unit　　　　Actor(s)/Cameras/Movement/Properties

Lighting Instruments
　Key Lights:
　Fill Lights:
　Soft Lights:

Lighting Accessories
　Barn Doors:
　Flags:
　Gobo:

Filters
　Spun Glass:
　Gels:
　Scrims:

Windows
　Daylight:
　Night time:
　Dusk:
　Change:

Property Lights
　Lamps:
　Ceiling:
　Other:

In-Cue　　　　　Dialogue/Action
　　　　　　Lighting Change Cue

Out-Cue　　　　Dialogue/Action
　　　　　　Lighting Change Cue

Comments:

EFFECTS/MUSIC BREAKDOWN

MULTIPLE CAMERA VIDEO DRAMA PRODUCTION

Audio Director: Teleplay Title:

Producer: Director: Author:

Approval ☐   Approval ☐   Date: / /   Page   of

SCRIPT UNIT/PAGE/LINE	In-cue/Out-cue	Effect	Est. Length	Music
			:	
			:	
			:	
			:	
			:	
			:	
			:	
			:	
			:	
			:	
			:	
			:	
			:	
			:	
			:	
			:	
			:	
			:	

Rights/Clearances Required:

    Effect/Music: Publisher: Copyright Owner:

    Effect/Music: Publisher: Copyright Owner:

SCRIPT UNIT/PAGE/LINE	In-cue/Out-cue	Effect	Est. Length	Music
			:	
			:	
			:	
			:	
			:	
			:	
			:	
			:	
			:	
			:	
			:	
			:	
			:	
			:	
			:	
			:	
			:	
			:	
			:	
			:	
			:	
			:	

Effects/Music Breakdown (Continued) Page   of

AUDIO PLOT

MULTIPLE CAMERA VIDEO DRAMA PRODUCTION

Audio Director:

Producer:

Director:

Teleplay Title:

Author:

Approval ☐

Approval ☐

Script Page ☐

Script Unit ☐

Set:

Date: /   /

Page   of

Lighting
☐ Interior
☐ Exterior
☐ Day
☐ Night

Sound
☐ Synchronous
☐ Silent
☐ _____

Microphone
☐ Directional
☐ Wireless
☐ _____

Microphone Support
☐ Fishpole
☐ Giraffe
☐ Handheld
☐ Hanging
☐ _____

Sound Effects
☐ Foldback
☐ _____

Bird's Eye Floor Plan

Cameras /Properties/Blocking
Microphone Grip(s)/Microphone Support(s)/Cable Run(s)
Sound Perspective: ☐ close   ☐ distant

Description of Take/Unit

Actor(s)/Cameras/Movement/Properties

In-Cue   Dialogue/Action

Out-Cue   Dialogue/Action

Comments:

AUDIO PICKUP PLOT

MULTIPLE CAMERA VIDEO DRAMA PRODUCTION

Mike Grip 1:

Mike Grip 2:

Audio Director:

       Approval ☐

Teleplay Title:

Author:

Date: / /

Page   of

SET: SCRIPT UNIT(S):

Bird's eye view of the set with set/properties/actor(s)/actress(es)/camera placement.
Indicate microphone grip/audio pickup placement. Placement may change for varying takes/actor/
actress movement, etc.

NOTES

PRODUCTION SCHEDULE

MULTIPLE CAMERA VIDEO DRAMA PRODUCTION

Director:	Teleplay Title:	
Producer:	Author:	
Production Facility:	Script Length:	pages
Date: / /	Script Units:	units

PRODUCTION DAY ONE   Date: /   /   Crew Call: :   Cast Call   :

UNIT	STUDIO TIME	SET	CAST	COSTUMES/ PROPERTIES	NOTES
	:				
	:				
	:				
	:				
	:				
	:				
	:				
	:				
	:				
	:				

PRODUCTION DAY TWO   Date: /   /   Crew Call: :   Cast Call   :

	:				
	:				
	:				
	:				
	:				
	:				
	:				
	:				
	:				
	:				
	:				

PRODUCTION DAY THREE Date: /   /   Crew Call: :   Cast Call   :

	:				
	:				
	:				
	:				
	:				
	:				
	:				
	:				
	:				
	:				
	:				

81

Production Schedule (Continued)					Page of

PRODUCTION DAY [ ] Date: / / Crew Call: : Cast Call: :

UNIT	STUDIO TIME	SET	CAST	COSTUMES/ PROPERTIES	NOTES
	:				
	:				
	:				
	:				
	:				
	:				
	:				
	:				
	:				
	:				
	:				

PRODUCTION DAY [ ] Date: / / Crew Call: : Cast Call: :

	:				
	:				
	:				
	:				
	:				
	:				
	:				
	:				
	:				
	:				
	:				
	:				

PRODUCTION DAY [ ] Date: / / Crew Call: : Cast Call: :

	:				
	:				
	:				
	:				
	:				
	:				
	:				
	:				
	:				

PRODUCTION DAY [ ] Date: / / Crew Call: : Cast Call: :

	:				
	:				
	:				
	:				
	:				
	:				

CAMERA SHOT LIST
MULTIPLE CAMERA VIDEO DRAMA PRODUCTION

Camera Operator:
Camera: C1 C2 C3
Studio Production Date: /   /
Set:

Teleplay Title:
Author:
Date: /   /
Script Unit ☐

Page   of

SHOT No.	CAMERA FRAMING	CHARACTER/ OBJECT	CAMERA MOVEMENT	SPECIAL INSTRUCTIONS

Camera Shot List (Continued)				Page of
SHOT No.	CAMERA FRAMING	CHARACTER/ OBJECT	CAMERA MOVEMENT	SPECIAL INSTRUCTIONS

CONTINUITY NOTES

MULTIPLE CAMERA VIDEO DRAMA PRODUCTION

Continuity Person:	Teleplay Title:
	Author:
	Taping Date: / / Page of

SET		INTERIOR	DAY	ENDING CAMERA C1 C2 C3	VIDEOTAPE No.
		EXTERIOR	NIGHT	ENDING SHOT No.	SCRIPT UNIT No.
		SCRIPT PAGE No.	PRECEDING EDIT UNIT No.		CUT ☐ DISSOLVE ☐ WIPE ☐
	DETAILS		SUCCEEDING EDIT UNIT No.		CUT ☐ DISSOLVE ☐ WIPE ☐

ACTOR(S)/ACTRESS(ES)/COSTUME/MAKE-UP/PROPERTIES NOTES

CIRCLE TAKES	1	2	3	4	5	6	7	8	9	10	11	12
END SLATE												
TIMER												
COUNTER												
REASON FOR USE/ NOT GOOD												

ACTION   DIALOGUE

VIDEOTAPE LOG

MULTIPLE CAMERA VIDEO DRAMA PRODUCTION

Director: _____
Assistant Director: _____
Log Form No. ☐

Teleplay Title: _____
Author: _____
Taping Date: / /   Page   of

Script Page	Script Unit	Videotape No.		Set		Notes					
CIRCLE TAKES	1	2	3	4	5	6	7	8	9	10	11	12
END SLATE												
TIMER												
COUNTER												
REASON FOR USE/ NOT GOOD												

Script Page	Script Unit	Videotape No.		Set		Notes					
CIRCLE TAKES	1	2	3	4	5	6	7	8	9	10	11	12
END SLATE												
TIMER												
COUNTER												
REASON FOR USE/ NOT GOOD												

Script Page	Script Unit	Videotape No.		Set		Notes					
CIRCLE TAKES	1	2	3	4	5	6	7	8	9	10	11	12
END SLATE												
TIMER												
COUNTER												
REASON FOR USE/ NOT GOOD												

Script Page	Script Unit	Videotape No.		Set		Notes						
CIRCLE TAKES	1	2	3	4	5	6	7	8	9	10	11	12
END SLATE												
TIMER												
COUNTER												
REASON FOR USE/ NOT GOOD												

Script Page	Script Unit	Videotape No.		Set		Notes						
CIRCLE TAKES	1	2	3	4	5	6	7	8	9	10	11	12
END SLATE												
TIMER												
COUNTER												
REASON FOR USE/ NOT GOOD												

Script Page	Script Unit	Videotape No.		Set		Notes						
CIRCLE TAKES	1	2	3	4	5	6	7	8	9	10	11	12
END SLATE												
TIMER												
COUNTER												
REASON FOR USE/ NOT GOOD												

TALENT RELEASE

MULTIPLE CAMERA VIDEO DRAMA PRODUCTION

Talent Name: _____

Actor/Actress/Extra   *(Please Print)*

Teleplay Title: _____

For value received and without further consideration, I hereby consent to the use of all photographs, videotapes, or film taken of me and/or recordings made of my voice and/or written extraction, in whole or in part, of such recordings or musical performance

at _____ on _____ 19___

    (Recording Location)   *(Month)*   *(Day)*   *(Year)*

by_____ for_____

    (Producer)   *(Producing Organization)*

and/or others with its consent, for the purposes of illustration, advertising, or publication in any manner.

Talent Name_____

         (Signature)

Address_____ City_____

State _____ Zip Code _____

Date:___/___/___

If the subject is a minor under the laws of the state where modeling, acting, or performing is done:

Guardian _____ Guardian _____

         (Signature)   *(Please Print)*

Address _____ City _____

State _____ Zip Code _____

Date:___/___/___

POSTPRODUCTION CUE SHEET

MULTIPLE CAMERA VIDEO DRAMA PRODUCTION

Producer:

Director:

Editor:

Teleplay Title:

Author:

Date: /   /

SCRIPT UNIT	LOG FORM	TAPE NO.	TAKE NO.	TIME/ CODE	LENGTH	COMMENTS	
				:	:		
				:	:		
				:	:		
				:	:		
				:	:		
				:	:		
				:	:		
				:	:		
				:	:		
				:	:		
				:	:		
				:	:		
				:	:		
				:	:		
				:	:		
				:	:		
				:	:		
				:	:		
				:	:		
				:	:		
				:	:		
				:	:		
				:	:		
				:	:		
				:	:		
				:	:		
				:	:		
				:	:		
				:	:		
				:	:		

SCRIPT UNIT	LOG FORM	TAPE NO.	TAKE NO.	TIME/ CODE	LENGTH	COMMENTS	
				:	:		
				:	:		
				:	:		
				:	:		
				:	:		
				:	:		
				:	:		
				:	:		
				:	:		
				:	:		
				:	:		
				:	:		
				:	:		
				:	:		
				:	:		
				:	:		
				:	:		
				:	:		
				:	:		
				:	:		
				:	:		
				:	:		
				:	:		
				:	:		
				:	:		
				:	:		
				:	:		
				:	:		
				:	:		
				:	:		
				:	:		
				:	:		
				:	:		
				:	:		
				:	:		
				:	:		
				:	:		
				:	:		
				:	:		

TITLING/CREDITS COPY

MULTIPLE CAMERA VIDEO DRAMA PRODUCTION

Production Assistant: | Date: / / | Page of

SLATE

Title: [   ]
Producer: [   ]
Director: [   ]
Date: /   /
Videotape No. [ ]
Script Unit No. [ ]
Take No. [ ]

Record No.

**PRODUCTION GROUP/
COMPANY**

[   ]

[   ]

PRESENTS

Font Style:
Font Size:
Color:
Effects:
Record No.:
Front Time: :
Dialogue Cue:

ORDER [ ]
Rolled ☐   Speed ☐
Advanced ☐ Time   :

TELEPLAY TITLE

[   ]

[   ]

Font Style:
Font Size:
Color:
Effects:
Record No.:
Front Time: :
Dialogue Cue:

ORDER [ ]
Rolled ☐   Speed ☐
Advanced ☐ Time   :

TELEPLAY AUTHOR

Written by

[   ]

Font Style:
Font Size:
Color:
Effects:
Record No.:
Front Time: :
Dialogue Cue:

ORDER [ ]
Rolled ☐   Speed ☐
Advanced ☐ Time   :

LEADING
ACTOR/ACTRESS

Starring

as

Font Style:
Font Size:
Color:
Effects:
Record No.:
Front Time:　　:
Dialogue Cue:

ORDER ☐
Rolled ☐　　Speed ☐
Advanced ☐　Time　　:

and

as

Font Style:
Font Size:
Color:
Effects:
Record No.:
Front Time:　　:
Dialogue Cue:

ORDER ☐
Rolled ☐　　Speed ☐
Advanced ☐　Time　　:

DIRECTOR

Directed by

Font Style:
Font Size:
Color:
Effects:
Record No.:
Front Time:　　:
Dialogue Cue:

ORDER ☐
Rolled ☐　　Speed ☐
Advanced ☐　Time　　:

PRODUCER

Produced by

Font Style:
Font Size:
Color:
Effects:
Record No.:
Front Time:　　:
Dialogue Cue:

ORDER ☐
Rolled ☐　　Speed ☐
Advanced ☐　Time　　:

Font Style:
Font Size:
Color:
Effects:
Record No.:
Front Time: :
Dialogue Cue:

ORDER ☐
Rolled ☐   Speed ☐
Advanced ☐ Time   :

Font Style:
Font Size:
Color:
Effects:
Record No.:
Front Time: :
Dialogue Cue:

ORDER ☐
Rolled ☐   Speed ☐
Advanced ☐ Time   :

Font Style:
Font Size:
Color:
Effects:
Record No.:
Front Time: :
Dialogue Cue:

ORDER ☐
Rolled ☐   Speed ☐
Advanced ☐ Time   :

Font Style:
Font Size:
Color:
Effects:
Record No.:
Front Time: :
Dialogue Cue:

ORDER ☐
Rolled ☐   Speed ☐
Advanced ☐ Time   :

CLOSING CREDITS

Technical Director

Videotape Editor

Assistant Director

Production Assistant

Font Style:
Font Size:
Color:
Effects:
Record No.:
Front Time: :
Dialogue Cue:

ORDER ☐
Rolled ☐   Speed ☐
Advanced ☐ Time   :

Audio Director

Microphone Grips

Font Style:
Font Size:
Color:
Effects:
Record No.:
Front Time: :
Dialogue Cue:

ORDER ☐
Rolled ☐   Speed ☐
Advanced ☐ Time   :

Studio Camera Operators

Floor Director

Font Style:
Font Size:
Color:
Effects:
Record No.:
Front Time: :
Dialogue Cue:

ORDER ☐
Rolled ☐   Speed ☐
Advanced ☐ Time   :

Video Engineer

Videotape Recording

Lighting Director

Font Style:
Font Size:
Color:
Effects:
Record No.:
Front Time: :
Dialogue Cue:

ORDER ☐
Rolled ☐   Speed ☐
Advanced ☐ Time   :

CLOSING CREDITS

Cast

ORDER ☐
Rolled ☐   Speed ☐
Advanced ☐   Time   :

Font Style:
Font Size:
Color:
Effects:
Record No.:
Front Time: :
Dialogue Cue:

Set Designer

Properties Master/Mistress

Make-Up Artist

Costume Master/Mistress

ORDER ☐
Rolled ☐   Speed ☐
Advanced ☐   Time   :

Font Style:
Font Size:
Color:
Effects:
Record No.:
Front Time: :
Dialogue Cue:

Copyright 1991

ORDER ☐
Rolled ☐   Speed ☐
Advanced ☐   Time   :

Font Style:
Font Size:
Color:
Effects:
Record No.:
Front Time: :
Dialogue Cue:

ORDER ☐
Rolled ☐   Speed ☐
Advanced ☐   Time   :

Font Style:
Font Size:
Color:
Effects:
Record No.:
Front Time: :
Dialogue Cue:

EFFECTS/MUSIC CUE SHEET

MULTIPLE CAMERA VIDEO DRAMA PRODUCTION

Audio Director:

Director/Editor:

Teleplay Title:
Author:
Date: /   /
Page   of

SCRIPT UNIT	EFFECT	MUSIC	SOURCE		LENGTH	IN-CUE / OUT-CUE
						IN-CUE
						OUT-CUE
			CUT☐ ☐CUT			IN-CUE
			FADE☐ ☐FADE		:	OUT-CUE
			CUT☐ ☐CUT			IN-CUE
			FADE☐ ☐FADE		:	OUT-CUE
			CUT☐ ☐CUT			IN-CUE
			FADE☐ ☐FADE		:	OUT-CUE
			CUT☐ ☐CUT			IN-CUE
			FADE☐ ☐FADE		:	OUT-CUE
			CUT☐ ☐CUT			IN-CUE
			FADE☐ ☐FADE		:	OUT-CUE
			CUT☐ ☐CUT			IN-CUE
			FADE☐ ☐FADE		:	OUT-CUE
			CUT☐ ☐CUT			IN-CUE
			FADE☐ ☐FADE		:	OUT-CUE
			CUT☐ ☐CUT			IN-CUE
			FADE☐ ☐FADE		:	OUT-CUE
			CUT☐ ☐CUT			IN-CUE
			FADE☐ ☐FADE		:	OUT-CUE
			CUT☐ ☐CUT			IN-CUE
			FADE☐ ☐FADE		:	OUT-CUE
			CUT☐ ☐CUT			IN-CUE
			FADE☐ ☐FADE		:	OUT-CUE
			CUT☐ ☐CUT			IN-CUE
			FADE☐ ☐FADE		:	OUT-CUE
			CUT☐ ☐CUT			IN-CUE
			FADE☐ ☐FADE		:	OUT-CUE
			CUT☐ ☐CUT			IN-CUE
			FADE☐ ☐FADE		:	OUT-CUE
			CUT☐ ☐CUT			IN-CUE
			FADE☐ ☐FADE		:	OUT-CUE
			CUT☐ ☐CUT			IN-CUE
			FADE☐ ☐FADE		:	OUT-CUE
			CUT☐ ☐CUT			IN-CUE
			FADE☐ ☐FADE		:	OUT-CUE

SCRIPT UNIT	EFFECT	MUSIC	SOURCE		LENGTH	IN-CUE
						OUT-CUE
			CUT☐	☐CUT		IN-CUE
			FADE☐	☐FADE	:	OUT-CUE
			CUT☐	☐CUT		IN-CUE
			FADE☐	☐FADE	:	OUT-CUE
			CUT☐	☐CUT		IN-CUE
			FADE☐	☐FADE	:	OUT-CUE
			CUT☐	☐CUT		IN-CUE
			FADE☐	☐FADE	:	OUT-CUE
			CUT☐	☐CUT		IN-CUE
			FADE☐	☐FADE	:	OUT-CUE
			CUT☐	☐CUT		IN-CUE
			FADE☐	☐FADE	:	OUT-CUE
			CUT☐	☐CUT		IN-CUE
			FADE☐	☐FADE	:	OUT-CUE
			CUT☐	☐CUT		IN-CUE
			FADE☐	☐FADE	:	OUT-CUE
			CUT☐	☐CUT		IN-CUE
			FADE☐	☐FADE	:	OUT-CUE
			CUT☐	☐CUT		IN-CUE
			FADE☐	☐FADE	:	OUT-CUE
			CUT☐	☐CUT		IN-CUE
			FADE☐	☐FADE	:	OUT-CUE
			CUT☐	☐CUT		IN-CUE
			FADE☐	☐FADE	:	OUT-CUE
			CUT☐	☐CUT		IN-CUE
			FADE☐	☐FADE	:	OUT-CUE
			CUT☐	☐CUT		IN-CUE
			FADE☐	☐FADE	:	OUT-CUE
			CUT☐	☐CUT		IN-CUE
			FADE☐	☐FADE	:	OUT-CUE
			CUT☐	☐CUT		IN-CUE
			FADE☐	☐FADE	:	OUT-CUE
			CUT☐	☐CUT		IN-CUE
			FADE☐	☐FADE	:	OUT-CUE
			CUT☐	☐CUT		IN-CUE
			FADE☐	☐FADE	:	OUT-CUE
			CUT☐	☐CUT		IN-CUE
			FADE☐	☐FADE	:	OUT-CUE

Chapter 4

Description
and Glossary
for Studio Production
Organizing Forms

INTRODUCTION

A barometer to successful television production is the degree to which producing and production tasks are accomplished and the extent to which producing and production tasks are *successfully* accomplished. Equally important to successful accomplishment of those tasks is the knowledge and awareness of the details necessary at each producing and production stage.

This chapter presents a description and glossary of terms for successful studio drama production task accomplishment. These descriptions and glossaries are presented in terms of the organizing forms found in Chapter 3. Each organizing form is presented here by title, the particular production process to which it refers, the producing or production personnel responsible for the completion of the form and the tasks, and a description of the purpose of the form and of the objective in using the form.

Not all organizing forms are necessary or even required in every studio drama production. They are selective and designed to be used as needed. Those forms that organize and assist in accomplishing a studio drama preproduction, production, or postproduction task should be tried for the ease and thoroughness with which they organize a task. Those forms that are redundant to a particular drama production should be ignored. The forms are meant to be helpful, not a stumbling block, to studio drama production.

DESCRIPTION AND GLOSSARY

- **Production Budget Form**

Production process Studio drama preproduction

Responsibility Producer

Purpose To realistically estimate all possible costs of the proposed studio drama from preproduction to postproduction.

Objective The budget form is a blank model of a budget for the production of a studio drama. The form is meant to organize as many facets of multiple camera drama production as can be anticipated into estimated expenses and, after production, into actual expenses. The form suggests possible cost items across the spectrum of studio video production, personnel, equipment, labor, time, and materials.

The form should be used to suggest expenses and to call attention to possible hidden costs before production begins.

General comments on the use of the budget form
The form should be studied for line items that might pertain to a proposed multiple camera drama production. Only applicable line items need to be considered. Use the suggestion of line items to consider as many foreseeable real costs as possible for the project.

Note that all sections of the budget are summarized on the front page of the form. The subtotals of costs from the individual sections are brought forward to be listed in the summary.

Some costs are calculated by the number of days employed, the hourly rate of pay, and overtime hours. Other costs are calculated by the number of items or people and the allotted amount of money per day for the item per person. The cost of materials is calculated by the amount of materials times the cost per unit.

Most cost entry columns of this budget form are labeled "Days, Rate, O/T [overtime] Hrs, Total." For those line items that do not involve day and rate, the total column alone should be used.

Projected costs to be applied to the budget can be determined from a number of sources. One source of equipment rental and production costs is the rate card of a local video production facility. Salaries and talent fees can be estimated from the going rates of relevant services of equivalent professionals or from their agencies. Cost quotations can be requested in phone calls to providers of services and materials. Travel costs can be obtained by a phone call to a travel agency or an airline company. Preparing a good budget is going to involve a lot of time and research.

Glossary

Producer The name of the producer for the teleplay is entered here.

Director The name of the director of the drama is recorded in this space. If a director has not yet been chosen or hired, the space can be left blank.

Date The date of the preparation of the studio drama production budget is entered here.

Teleplay Title The title for the proposed studio drama should be entered here.

Author The name of the author of the teleplay being proposed should be entered here.

Length The length of the proposed studio drama should be recorded here.

No. Preproduction Days/Hours Estimate the number of days and hours it will take to complete the necessary preproduction stages for the drama. This estimate includes time on all preproduction stages (e.g., budget, production schedule, script breakdown, casting, master script, and shot list) for all crew members involved in the preproduction of the drama (e.g., producer, director, camera operators, lighting director, videotape recorder operator, audio director, continuity person, and talent).

No. Studio Production Days/Hours Estimate the number of days and hours for which the studio or control room facilities may be needed in the production of the drama. This time estimate accounts for all in-studio videotaping (e.g., talent on a set) and the use of the control room switcher or character generator for production.

No. Postproduction Days/Hours Estimate the number of days and hours that will be spent in postproduction in completing the master edit.

First Production Date Indicate the date proposed to begin studio production and videotaping of the drama.

Completion Date Indicate the projected date of completion of the drama. This date should include all postproduction tasks.

Summary of Production Costs This summary area brings forward the subtotals of costs from the respective sections within the budget form. Note that each item in lines 1 through 19 is found as a section head within the budget form. When each section is completed, the section subtotal is listed in this summary table.

Contingency This term refers to the practice of adding a percentage to the subtotal of the summary section (i.e., to lines 1 through 19) as a pad against the difference between actual costs and the estimated costs of all line items. The customary contingency percentage

is 15%. Multiply the subtotal in line 20 by 0.15 to determine the contingency figure. Enter that result into line 21. The total of lines 20 and 21 becomes the grand total of the budget costs.

Estimated The *estimated* category refers to costs that can only be projected before the production. Rarely are the estimated costs the same as the final costs because of the number of variables that cannot be anticipated. Every attempt should be made to ensure realistic estimated cost figures and the consideration of as many variables as possible. The challenge of preparing an estimated budget is to project as close as possible to the actual costs.

Actual The *actual* category records the costs that are finally incurred for the respective line items during or after the production is completed. The actual costs are those costs that will really be paid. The ideal budget seeks to have actual costs come as close as possible to— i.e., be at or, preferably, under—the estimated costs.

Story Rights, Other Rights and Clearances This section accounts for the costs that might be incurred in obtaining necessary script story rights, copyright clearance, and music synchronization rights.

Teleplay Script This section suggests some of the possible costs that might be incurred with the writing or rewriting of the teleplay script for the drama production. On some story lines, searches may have to be made for legal purposes (e.g., authenticity for true life based stories or libel avoidance).

Producer and Staff This section accounts for the producer and other personnel that may be involved in producing responsibilities. The role of producer is divided into the three areas of a production: preproduction, production, and postproduction.

Director and Staff This section covers the director and personnel associated with the directing responsibilities. As with most budgets, some line items (e.g., secretary) are listed for purposes of suggesting enlarged staffs and additional roles that might be needed during multiple camera drama production. Not all of the line items are necessary to all studio drama productions.

Talent The expenses of any talent are accounted for in this section. Some talent may be hired by contract and may be subject to other fees (e.g., Screen Actors Guild (SAG)). Other talent may be freelancers. Space in the budget allows for listing the characters from the teleplay and the actor or actress playing the part.

Benefits Depending on the arrangements made with the talent, some production budgets may have to account for benefits accruing to the talent and crew members (e.g., health benefits or pensions). Insurance coverage and state and federal taxes that are due to any salaries to talent and crew members should be accounted for in this section.

Production Facility This section lists production facility space that may be required and, for each facility space, the technical hardware that may or may not be needed. Rate cards for production facilities provide cost information for these spaces and for the technical hardware in them. The producer will have to choose what is needed and determine the projected cost of each item.

Production Staff This section accounts for production staff and crew members for each stage of the production: preproduction, production, and postproduction.

Cameras, Videotape Recording, and Video Engineering Those personnel and expenses associated with the maintenance and use of the video camera are accounted for in this section. This section records the equipment that will need to be purchased or rented. Camera equipment rental will be necessary if any exterior stock shots have to be made.

Set Design, Construction, and Decoration Studio drama production will involve the use of studio set(s) (even the cyclorama is considered a set); the costs for set design, construction, and decoration must be accounted for in the budget. This section of the budget suggests some design and construction personnel and some construction and decoration costs.

Properties Most studio dramas require set properties, action properties (e.g., vehicles and animals), or hand properties (e.g., food). This section of the budget suggests and accounts for personnel and action properties, their care and handling for the production, and the same for hand properties.

Set Lighting The need to create mood and setting on sets requires the personnel and equipment for set lighting. Studio costs can be incurred with the use of light instruments. The light instruments necessary to light the set must be listed in the budget.

Audio Production Audio production and recording are essential to drama production. This section suggests personnel, hardware, microphones, and the need for equipment rental and purchase line items for the production.

Costuming Drama production requires costumes. The personnel, materials, production, and care of costumes are accounted for in this section.

Make-up and Hairstyling Make-up and hair preparation are necessary to drama production. This section of the budget suggests line items to be accounted for when make-up, hairstyling, and supplies are to be used.

Videotape Stock The provision of necessary videotape stock in required sizes is accounted for in this section.

Postproduction Editing The personnel, facilities, and materials for postproduction editing are accounted for in this section.

Music Most drama production will involve original or recorded music. This section of the budget suggests some of the potential costs and personnel involved in the music.

Photography It is not uncommon that photography will be part of the production of any teleplay. The need for still photographs of the studio set and talent for print advertising or publicity purposes is suggested here.

• **Facility Request Form**

Production process Studio drama preproduction

Responsibility Producer

Purpose To organize and request television production space and technical hardware from a production facility.

Objective The facility request form allows the producer to choose from among all possible available television production facilities and television production hardware for the teleplay's production. The form prompts the producer to facility space and equipment, preproduction requirements, and production crew.

Glossary

Production Facility The name or title of the production facility at which the teleplay will be videotaped is entered here.

Producer The name of the producer of the drama is noted here.

Director The name of the director of the drama is entered here.

Date The date on which the facility request form was created is logged here.

Teleplay Title The title of the teleplay is listed here.

Teleplay Length The expected telecast length of the teleplay is indicated here.

Production Dates/Hours The range of dates necessary for the production of the teleplay is entered here. The hours of the days requested are also indicated.

Facility Space The producer chooses from among the possible facility spaces those that are required.

Studio Requirements The producer checks those elements that will be required in the production of the teleplay. In choosing the light instruments needed, the producer should indicate the number of such instruments needed and the required kilowatts of the instruments.

Control Room Requirements The producer chooses the equipment in the control room that will be needed.

Master Control Area Requirements The equipment needed from master control is checked here. If more than one unit is required, the number of units should be entered instead of a check mark.

Audio Control Room Requirements The producer must check the equipment from the available audio control equipment that will be required for the teleplay. As with other choices, if more than one unit is required, the number of units should be entered in the box instead of a check mark.

Preproduction Requirements The producer must indicate to the production facility what preproduction time will be required. Preproduction requirements include time for set construction, set lighting, and rehearsals.

Studio Production Personnel The producer must check with the production facility management which production crew personnel will be supplied by the facility and which will have to be supplied by the producing staff. This area of the form permits listing the names of the crew members and checking from which source they are coming.

• **Talent Audition Form**

Production process Studio drama preproduction

Responsibility Producer, director

Purpose To serve as an application form for prospective actors and actresses who may wish to audition for a part in the cast.

Objective The talent audition form elicits from an auditioning actor or actress all of the applicable information that the producer and director may need as they evaluate the videotaped performance of the prospective cast member. The form records personal information from the applicant that will serve such preproduction stages as costume construction and actor/actress availability. The form also requires a resumé and photograph of the applicant.

Glossary

Teleplay Title/Author This form requires the teleplay designation by title and author.

Actor/Actress The prospective cast member places his or her name in this space.

Telephone (Home/Work)/Address/City/State/Zip Code/ Birthdate This section collects relevant personal information on the applicant.

Availability/Unavailability This space elicits necessary information on whether the applicant can be available for the required preproduction and production dates.

Personal Data: Height/Weight/Ethnicity/Hat Size/Shoe Size/Waist Measurement/Chest Measurement/In Seam/Shirt or Blouse Size/Suit or Dress Size This block of information records all possible data on the applicant, which will serve costuming and make-up as background should the applicant be cast in the teleplay.

Agent/Union Affiliation This information defines professional applicants who are affiliated with a casting agent or an actors' union. Should the producer and director choose a union or agent affiliated cast member, costs may be incurred or expected salary requirements made.

Questions Four self-explanatory questions are asked of the applicant that will assist the producer and director in evaluating the prospective cast member.

Resumé/Photo/Experience/Training This information, which the applicant either attaches if the applicant has a resumé and photo) or lists (previous experience or training), will help the producer and director judge the applicant.

- **Characterization Form**

Production process Studio drama preproduction

Responsibility Producer, director, cast

Purpose To structure the initial process of developing a character for an actor or actress and the whole cast.

Objective The characterization form asks meaningful questions of each cast member, which should serve to define and develop the character required of the actor or actress. By triggering written responses from each cast member, most facets of a character are listed and probed.

Glossary

Director/Actor/Actress This form may be used by the director in assisting each cast member or by each actor or actress. The name of the director, actor, or actress is placed in this space.

Teleplay Title/Author The title and author of the teleplay are listed here.

Date The date on which the characterization form was completed is placed here.

Character The name of the character from the teleplay who must be characterized by an actor or actress is listed here.

Descriptive Adjectives After reading the script, the director, actor, or actress is asked to list adjectives that describe the character being developed.

Order of Importance After listing descriptive adjectives, the adjectives should be listed in the order of importance perceived by the cast member or the director.

Order Revealed to the Audience The adjectives should be listed in the order in which the cast member or director perceives them to be revealed to the audience.

Actions Performed by and Done to the Character This section requires an analysis of the actions performed by the character and those actions done to the character.

Action/Significance The actions performed by and done to the character are listed in the action column; the significance of each action is indicated in the second column.

Major Motivations of the Character The major motivations—those things that influence the character—are listed by script unit.

Character's Relationship to All Other Characters The relationship of the character to all other characters in the teleplay is listed.

Function the Character Plays in the Teleplay The function the character plays in the teleplay is labeled and elaborated here.

Audience's Emotional Reaction to the Character during the Teleplay The cast member or director describes what the audience's reaction should be to the character at important moments during the teleplay.

Change in the Character during the Teleplay The cast member or director should describe the change that takes place in the character from the beginning to the end of the teleplay.

How to Convey the Changes to the Audience The cast member or director should indicate how the changes occurring in the character should be conveyed to the audience by the actor or actress.

Character History The last step in characterization is to develop a history, in autobiographical form, of the character before and after the period of the teleplay, from birth to death.

- **Script Breakdown Form**

Production process Studio drama preproduction

Responsibility Director

Purpose To break the script down into production units usually according to sets, properties, or cast requirements.

Objective The script breakdown form organizes the teleplay script by breaking down production units according to differing drama elements (e.g., set, cast, and time of day). The script breakdown form will help organize the shooting units by script pages or number of script lines and production values criteria for scheduling and managing the studio production.

Glossary

Producer/Director The names of the producer and director are listed here.

Teleplay Title/Author The title of the teleplay and its author are placed here.

Length This entry records the estimated length of the final teleplay production.

Script Length The total number of script pages is recorded here.

Page of This records the expected number of pages containing the script breakdown and the number of each individual page.

Script Unit Breaking the script down into script units is a way of organizing drama scripts for production. Script units replace the classic theatrical division of a stage script into acts and scenes; in teleplays, acts and scenes are converted into consecutively numbered units. A unit is a manageable portion of the teleplay by video production standards. The script unit being analyzed should be listed in this column.

Script Pages/Lines The number of pages or lines of script copy for a unit of the script is a determinant of the length of a shooting unit. The page count can be made in whole page and portion of page numbers. The line count sums every line of copy in the unit.

Int/Ext These abbreviations stand for ''interior/exterior'' and refer to the script demands for an indoor or outdoor setting. Whether the set required is an interior or an exterior set can be a determinant for the script breakdown and studio shooting unit.

Time The time of day or night and the interior/exterior setting requirements should be recorded here.

Set Set means the specific required interior or exterior site or environment for a shooting unit of the teleplay (e.g., the bedroom at the Smiths' house).

Properties In the properties column should be listed all the properties, whether set properties, hand properties, or action properties required during the respective script unit.

Cast This entry records by name the specific characters who are required on the set for a particular unit.

Shooting Order When the entire script is broken down into production units, the director can determine the specific units for each production session and the unit shooting order for the entire studio production.

• **Set Design Form**

Production process Studio drama preproduction

Responsibility Set designer

Purpose To provide the set designer with a form and uniform grids on which to design studio set(s) from both a bird's eye view and a front view of the proposed set.

Objective The set design form provides the set designer the grid on which to prepare the basic sketch of the proposed set(s) for the teleplay. The grids are constructed to permit both a bird's eye view (a view from above the set) and a front view (a horizontal view, from the front looking into the set).

Glossary

Set Designer The name of the set designer is listed here.

Producer/Director/Approval The names of the producer and director of the teleplay are listed here, and an indication of their approval of the set design(s) is noted.

Teleplay Title/Author The title of the teleplay and its author are placed here.

Date The date of the completed design is entered here.

Page of This designation indicates the expected number of set design forms and the number of the individual page.

Set The name or title of the set is indicated here.

Script Units The units of the script that use the set design are listed here.

Bird's Eye View The larger grid is used for the view of the set design from above—the bird's eye view. This view affords the designation of width and depth, angles and corners, and the placement of set properties (e.g., chairs and lamps).

21 Units × 34 Units The grid contains 21 units on the vertical and 34 units on the horizontal side. The set designer can assign any convenient uniform unit of measure to the square unit. For example, by assigning two feet per square, the designer has an area of 42 feet by 68 feet with which to work.

Front View The smaller grid is used for a frontal view of the proposed set. Placing the frontal view directly below the bird's eye view allows alignment of the set, point for point, from both viewpoints. The frontal view is seen on the horizontal.

12 Units × 34 Units This grid contains 12 units on the vertical and 34 units on the horizontal side. The set designer can assign any convenient uniform unit of measure to the square unit.

• **Costume Design Form**

Production process Studio drama preproduction

Responsibility Costume master/mistress

Purpose To provide a standard analysis and description of each costume required.

Objective The costume design form records the details of each costume required for the teleplay. The form prompts all elements of costume design and measurement that will be needed during rental, purchase of materials, or construction and fitting of every costume. The form requires a simple sketch of the costume.

Glossary

Costume Master/Mistress The name of the costume master/mistress is indicated here.

Producer/Director/Approval The names of the producer and director are entered here. Both the producer and director indicate approval of the costume design by signing or initialing their respective approval of the design. The producer's approval authorizes the necessary expenditures to create the costumes.

Teleplay Title/Author The title of the teleplay and its author are listed here.

Character The character for which a costume is being designed is listed here.

Script Unit(s) The script unit(s) from the teleplay script for which the costume is required are listed here.

Costume No. Each costume required for every character is numbered. The number can indicate the successive costume within the teleplay for the character. Further coding can serve a cost accounting function for budget purposes (e.g., JD−3 would be the code for John Doe's third costume).

Costume Description The costume description records relevant details of the costume.

Historical Period Historical period places the costume within the period of history of which it is a part.

Costume Type Costume type describes the use of the costume (e.g., evening formal or bed clothes).

Fabric Fabric describes the type of material of which the costume is made.

Color Scheme The color scheme indicates the particular color design of the set, the teleplay, and the mood created by the director into which the costume must fit. The matching or contrasting colors that will become part of the costume should be described.

Trim Trim indicates the material chosen as edging, an accessory to the primary material of the costume.

Costume Elements List This list records specific elements to be used as part of the costume or in addition to the costume. This would include such items as shoes, socks, underwear, or suspenders.

Costume Accessories This list records those decorative additions apart from the essential elements of a costume, such as jewelry, boutonniere, cuff links, or gloves.

Costume Sketch This area invites a simple descriptive sketch of the costume from shoulders to feet.

Notes This block should be used to include any additional elements of the costuming not particularly described previously.

- **Make-up Design Form**

Production process Studio drama preproduction

Responsibility Make-up artist

Purpose To record the individual creative elements of each character's make-up requirements.

Objective The make-up design form provides a list of all possible make-up design elements that could go into the make-up requirements of each character in the teleplay. The form records the appearance requirements of the character. The make-up artist can also sketch the make-up design on a generic face form.

Glossary

Make-up Artist The make-up artist's name is entered here.

Producer/Director/Approval The names of the producer and director are entered here. The make-up artist requires design approval from both the producer and director. Each signs or initials approval of the design. The producer's approval includes the authorization of expenses for the purchase of necessary make-up supplies.

Teleplay Title/Author The title and author of the teleplay are listed here.

Date The date of make-up design completion is noted here.

Page of This notation indicates the expected number of make-up designs and the number of each individual page.

Character/Actor/Actress These entries record the character from the teleplay and the actor or actress cast in the role.

Age/Sex/Complexion/Type These categories summarize requirements of the character.

Wig/Moustache/Beard These choices indicate the facial hair or head hair that may be required.

Make-up Elements/Type or Code/Notes This listing records all possible make-up supplies that a character's make-up design could require. The appropriate box should be checked and the type or code of supply indicated. The notes column permits further elaboration on the make-up supplies, including the amount of an element needed.

Required Make-up Changes This box records any changes of make-up that may be required of a character (e.g., aging).

Script Unit(s)/Make-up Changes The script unit(s) requiring the changes are noted here along with the kind of make-up change required.

Special Requirements This area provides space to make any further notations of special make-up requirements.

- **Properties Breakdown Form**

Production process Studio drama preproduction

Responsibility Properties master/mistress

Purpose To record from the script all required set, hand, and action properties.

Objective The properties breakdown form records by script unit all of the properties needed for the teleplay. Properties can be those elements of furniture to decorate and dress a set, the hand properties that actors and actresses need (e.g., a pen or a glass), and action properties (e.g., a bicycle or a dog). The properties master/mistress carefully reads the script to create the breakdown list.

Glossary

Properties Master/Mistress The name of the properties master/mistress is indicated here.

Producer/Director/Approval The names of the producer and director are entered here. The properties master/mistress requires the approval of the properties breakdown list by the producer and director. They indicate their approval by signing or initialing the form. The producer's approval authorizes the acquisition of the approved properties by construction, purchase, or rental.

Teleplay Title/Author The title and author of the teleplay are entered here.

Script Length The length of the teleplay in terms of pages is entered here.

Script Units The length of the teleplay in terms of production units is entered here.

Page of This notation indicates the expected number of pages of the properties breakdown list and the number of each individual page.

Date The date of preparation of the properties breakdown list is entered here.

Script Unit The properties master/mistress indicates each script unit requiring properties.

Script Pages/Lines As a further indication of the use of a required property, a script page number or script line count into a unit is indicated here.

Character(s) This notation indicates the character who has to use the required property.

Properties This column records the required properties.

Notes This space allows for further description or labeling of the required properties or their use.

Insurance Required/Property/Type of Insurance The properties master/mistress indicates what properties will require insurance coverage. The producer makes note of this requirement and is responsible to provide coverage.

• **Blocking Plot Form**

Production process Studio drama preproduction

Responsibility Director

Purpose To create the director's master script—the studio and control room document from which the cast, dialogue, properties, cameras, camera framing, and composition will be directed.

Objective This essential preproduction stage prepares the director for all details for directing the cast and crew while in the studio and the control room during production. From the approved set design(s) and the scale model of each set, the director can design all production elements. The blocking plot serves to record the blocking of actor(s)/actress(es), cameras, set(s), and the storyboard sketch of every camera frame proposed for videotaping.

Some notes on the creation of the blocking plot form Many copies of the blocking plot form may be needed. One bird's eye view blocking frame is contained on each page. Each page of the blocking plot form is designed to be inserted into the director's copy of the master script facing the corresponding page of dialogue.

The director should sketch the bird's eye view of the set from the set design form. The director should then position the actor(s)/actress(es) in the floor plan and indicate their blocking placement and movement. The cameras are sketched into place to achieve the shot(s) as proposed in the storyboard frames in the right-hand column of the master script.

Each bird's eye floor plan is drawn and inserted adjacent to the master script with the corresponding storyboard frames on the script. The bird's eye floor plan need not be redrawn each time if there are no major changes. Once blocking is indicated on the blocking plot, cameras are added as the storyboard frames are sketched in on the master script pages.

Glossary

Director The director's name is entered here.

Date The date that the blocking plot was completed is noted here.

Set The name or label of the set for the script unit(s) being blocked is entered here.

Teleplay Title/Author The teleplay title and author are written here.

Script Page The number of the script page to which the blocking plot refers is recorded here.

Script Unit The corresponding script unit for which the blocking plot is being created is recorded here.

Lighting The general lighting cue for the set for this script unit is recorded in this column.

Interior/Exterior/Day/Night These are the evident choices for lighting cues to the lighting director.

Sound The general sound pickup requirement is recorded here.

Synchronous/Silent The possible choices for sound are synchronous (if sound is to be picked up with the video on the set) and silent (if video only may be required and no sound recorded).

Bird's Eye Floor Plan This space is the area in which the bird's eye view of the set is sketched for blocking purposes.

Cameras/Properties/Blocking These are the elements to be entered on the bird's eye view of the set. Cameras and actor(s)/actress(es) must be blocked and set properties (e.g., furniture), which are important to the set, must be included on the bird's eye view.

Description of Take/Unit This space records any verbal description of the action of the take or script unit. Some blocking may be better described in words than can be shown with sketched blocking.

Actor(s)/Actress(es)/Cameras/Movement/Properties These are the elements of the blocked unit that are to be included in any description of the take or unit being blocked.

In-cue/Dialogue/Action The in-cue is the dialogue or stage direction action that begins the take or script unit.

Out-cue/Dialogue/Action The out-cue is the dialogue or stage direction action on which the take or script unit ends.

Comments This section provides more space in which the director can make notes on the blocking for the unit. All thoughts should be jotted down during preproduction lest they be glossed over or forgotten during production.

- **Lighting Plot Form**

Production process Studio drama preproduction

Responsibility Lighting director

Purpose To prepare and organize the lighting design of the set(s) for the production.

Objective The lighting plot form is designed to facilitate the lighting design for the lighting director for each studio set. The plot helps the lighting director preplan the placement of lighting instruments, kind of lighting, lighting control, and lighting design. When lighting is preplanned, lighting equipment needs are easily realized and provided.

Notes on the use of the lighting plot form The lighting plot form is created to encourage preplanning for lighting design and production. Hence, the more lighting needs and aesthetics that can be anticipated, the better the lighting production tasks. At a minimum, every set lighting design should be created in advance of the production sessions. Lighting design can begin after the master script is complete. Essential to planning lighting design are having the bird's eye floor plan; the blocking of the actor(s)/actress(es); and information on whether the set is an interior or exterior setting, the time of day, and the mood of the script.

A different plot is not required for every script unit videotaped in the same set. Base lighting design need not change as script units within a set change. However, set changes will require new lighting set-up and design. It is these changes for which the lighting director must prepare.

Glossary

Lighting Director The name of the lighting director for the production is entered here.

Producer/Director/Approval The names of the producer and director are entered here. The lighting plots must be approved by the producer and director. They should sign or initial their approval on this form. Approval by the producer authorizes expenses for lighting needs acquisition.

Set The lighting director indicates the name or label for the set being lighted.

Teleplay Title/Author The teleplay's title and author are recorded here.

Script Page The lighting director will require an orientation to the corresponding page(s) of the script for every set to be lighted.

Script Unit The script unit being produced is another orientation to the script to benefit the lighting director.

Date The date of the completion of the lighting plot should be entered here.

Page of Each lighting plot page should be numbered consecutively and each page recorded with the respective page number as the portion of the whole, e.g., "Page 1 of 1," or "Page 2 of 5."

Lighting: Interior/Exterior/Day/Night This area of the plot notes the basic options for lighting design: inside or outside, day or night.

Lighting Change: Yes/No It is important to lighting design and lighting control to know if the script requires or the director plans any change of lighting during the unit being videotaped. This notation alerts the lighting director to that need.

Bird's Eye Floor Plan The best preproduction information for the lighting director is the floor plan for the set with actor(s)/actress(es) and camera blocking and movement indicated. This is sketched in this box from the master script. Where the cameras are to be placed is also important to the lighting director and lighting design.

Description of Take/Unit Any director's notes on the elements of the take that will affect lighting design should be noted here.

Lighting Instruments/Lighting Accessories/Filters/Windows/Property Lights These lists are designed to assist the lighting director in considering all elements of lighting design and materials or situations in preparing the lighting plot. The lighting director can make notations in the proper spaces in planning for the design of each set.

In-cue The lighting director may find it convenient to note the in-cue from the script when—on what action or dialogue—the lighting is to begin.

Out-cue The same notation on an out-cue of action or dialogue for the end of a shot or a change of lighting may be advantageous to the lighting director.

- **Effects/Music Breakdown Form**

Production process Studio drama preproduction

Responsibility Audio director

Purpose To provide a listing of all of the required sound effects and music for the teleplay and to note any rights or clearances needed.

Objective The effects/music breakdown form serves the audio director as the script breakdown form serves the director. The audio director studies the script for all required and suggested sound effects and music. Suggested sound effects and music are the creative choices of the audio director.

Glossary

Audio Director The name of the audio director is listed here.

Producer/Director/Approval The names of the producer and director are entered here. The audio director must submit this form for the approval of the producer and director. Approval by the producer authorizes expenses for the acquisition of needed effects and music sources.

Teleplay Title/Author The title and author of the teleplay are entered here.

Date The date of preparation of the breakdown form is entered here.

Page of This notation indicates the expected number of pages of the breakdown and the number of any single page.

Script/Unit/Page/Line The audio director records by the production unit, the teleplay page number, or numbered line on a designated page the places in the teleplay script where any effect or music is required or suggested.

In-cue/Out-cue The audio director records the corresponding in-cue or out-cue from the script dialogue or stage directions for the beginning and the end of any effect or music on the appropriate pages.

Effect This column records the particular sound effect required or designed for the cues noted.

Est. Length The portion of either the required or designated effect or music is the estimated length or duration of the effect or the music. That information is recorded here.

Music This column records any required or designated music for the cues noted.

Rights/Clearances Required The audio director notes here any rights or clearances that will have to be secured before the effects or music can be used.

- **Audio Plot Form**

Production process Studio drama preproduction

Responsibility Audio director

Purpose To facilitate and encourage the design of sound perspective and recording for studio drama production.

Objective The audio plot form is designed to prompt the audio director to weigh the set environment and sound production values in planning for audio equipment and quality microphone pickup and recording during studio drama production.

Notes on the use of the audio plot form The audio plot form is intended to encourage preproduction by the audio director. One plot for every set is probably adequate, although multiple forms might be required for a very detailed and complicated sound recording take.

Glossary

Audio Director The name of the audio director is placed here.

Producer/Director/Approval The names of the producer and director are entered here. The audio director is required to get the approval of the producer and director for the audio plot. The approval of the producer authorizes the expenses to be incurred for audio coverage of the production.

Set The set for which the audio plot was designed is listed here.

Teleplay Title/Author The title of the teleplay and its author are listed here.

Script Page The audio director requires an orientation to the corresponding page(s) of a script for every set to be covered for sound.

Date The date the plot was designed is entered here.

Script Unit The script unit being videotaped is another orientation to the script to benefit the audio director.

Page of This notation indicates the expected number of pages of audio plot forms that have been designed and each individual page number.

Lighting: Interior/Exterior/Day/Night This area of the plot notes the basic options for sound perspective design: inside or outside, day or night.

Sound: Synchronous/Silent This choice of sound recording needs indicates what would be required on the set by the audio director. Synchronous sound indicates that both audio and video will be recorded during a take; silent means that video only is required.

Microphone: Directional/Wireless/Other The audio director can make a choice of microphones to be used during videotaping the proposed script unit.

Microphone Support: Fishpole/Giraffe/Handheld/Hanging/Other The audio director can indicate the proposed method of microphone support for sound coverage during production.

Sound Effects: Foldback/Other The audio director can indicate the required effects, including foldback to the cast and microphone boom grip(s)/operator(s).

Bird's Eye Floor Plan The best preproduction information for the audio director is the bird's eye floor plan with the blocking and movement of actor(s)/actress(es) indicated. This should be sketched in this box from the master script. Placement of the cameras is also important information for the audio director in the planning of sound design and microphone and microphone boom and grip placement.

Sound Perspective: Close/Distant When a change in framing is indicated it could trigger a change in sound perspective; the change in framing should be calculated into the basic sound design. The new framing composition is indicated by checking the appropriate box for new framing. Quality sound perspective creates the auditory sense that when an actor/actress is framed closely, sound levels should be higher; when an actor/actress is framed at a distance, sound levels should be lower. Close or distant framing will also indicate to an audio director that use of a microphone mount will have to be controlled to avoid catching the microphone or microphone or boom shadows in the camera shot.

Description of Take/Unit Any of the director's notes on the elements of the take that will affect sound design and recording should be noted here. For example, excessive movement of actor(s)/actress(es) or properties and expressive gestures could affect sound recording and control on a set. The fact that some sound playback may be required during a take would be noted here.

In-cue The in-cue, either from the dialogue or action, can affect the sound design of a take. That in-cue should be noted here.

Out-cue The out-cue or final words of dialogue or action can affect sound design. Notation of that out-cue should be made here.

- **Audio Pickup Plot Form**

Production process Studio drama preproduction

Responsibility Microphone boom grip(s)/operator(s)

Purpose To analyze and design proposed sound pickup placement on studio sets before production.

Objective The sound pickup plot form is an attempt to create as much quality preproduction analysis as possible from the production crew to benefit the production of the teleplay. With this form, the microphone boom grip(s)/operator(s) can take what preproduction information is available to the grip(s)/operator(s) and plan microphone and operator placement within a set before a production session. By studying the blocking of the actor(s)/actress(es) and the placement of cameras, cable runs and grip/operator placement can be proposed.

Glossary

Microphone Boom Grip/Operator 1 and 2 These entries list the microphone boom grip(s)/operator(s) assigned to the studio production. It may be that the analysis required for this plot will indicate that more than one operator or grip will be required for the production.

Audio Director/Approval The name of the audio director is entered here. The microphone boom grip(s)/operator(s) will need the audio director's approval for this plot.

Teleplay Title/Author The title and author of the teleplay are entered here.

Date The date of the preparation of the plot is indicated here.

Page of This notation indicates the expected number of plots that are part of the proposed sound coverage and the page number of each individual form.

Set The set for which the pickup plot is designed is labeled here.

Script Unit(s) The script unit or units for which the set and pickup plot is proposed are recorded here.

Bird's Eye View The bird's eye view of the set with the set itself, set properties, cast, and cameras in place is sketched here. This information is available from the director's master script. The blocking indicated will suggest to the microphone boom grip(s)/operator(s) what areas will have to be covered by the microphone(s) under the control of the grip(s)/operator(s). The microphone boom grip(s)/operator(s) are sketched into the set.

Notes This area of the audio plot form permits the microphone boom grip(s)/operator(s) to record any other elements or questions on sound coverage before production.

• **Production Schedule Form**

Production process Studio drama preproduction

Responsibility Director

Purpose To preplan the script units to be produced for every studio production session.

Objective The production schedule form organizes all elements of the drama for every production session. The director has to plan the number of script units, set availability, cast availability, and approximate times for each production session in the studio. All producing and production staff will need copies of this schedule in advance of the first scheduled production session.

Glossary

Director/Producer The names of the director and producer are listed here.

Production Facility The name of the production facility is entered here.

Date The date the production schedule was created is noted here.

Teleplay Title/Author The title and author of the teleplay are entered here.

Script Length This indication of the length of the script in pages confirms information already held by the cast and crew with scripts.

Script Units This indication of the number of total units in the script is also a confirmation of the extent of the script.

Production Day One The production schedule form is broken into any number of available production session days.

Date The date of each production day is confirmed with this entry.

Crew Call The director indicates the time of the required rendezvous for the production crew. The crew call may not necessarily be at the same time as the cast call. The studio crew, with responsibilities for major production equipment, may need more lead time than does the cast before beginning the first unit blocking.

Cast Call The director indicates the time of the required rendezvous of the cast members and staff. The cast call may not necessarily be at the same time as the crew call.

Unit This column lists the exact script unit(s) that the director plans to videotape during the studio production session.

Studio Time In this entry, the director plans the time needed to accomplish each scheduled script unit.

Set The director confirms the studio set(s) required for the scheduled script units.

Cast The cast that will be required for the script units scheduled is listed here.

Costumes/Properties The director indicates the required costumes and properties (or specialized properties) required for the scheduled script units.

Notes The director can indicate to the staff, cast, and crew any additional information relevant for the particular production session or script unit.

• **Camera Shot List Form**

Production process Studio drama preproduction

Responsibility Technical director

Purpose To translate the proposed storyboard frames from the master script into a shot list for each camera.

Objective The shot list form organizes each proposed storyboard frame on the director's master script into shots listed by studio camera. Each shot generates the necessary video to create the proposed edits as designed on the master script. The shot list form will translate each proposed shot from the master script to individual lists for the camera operators.

Glossary

Camera Operator The name of the camera operator is entered on this line.

Camera: C1, C2, C3 The camera assigned to the particular camera operator is indicated here.

Studio Production Date The date of the production session scheduled for the listed camera shots is entered here.

Set The set being used for the listed camera shots is recorded here.

Teleplay Title/Author The title of the teleplay and its author are listed here.

Date The date of the completion of the shot list form should be entered here.

Script Unit The script unit for which the camera shot list is being prepared is recorded here. The technical director produces separate shot lists for each camera and for each script unit.

Shot No. Each proposed shot needed to create the storyboard frames of the script is numbered consecutively on the director's master script. The shot numbers are camera specific. The technical director separates the numbered shots from the master script and lists them for each camera in the order in which they appeared in the master script. Each shot list will contain the numbered shots assigned to the camera for which the list was prepared. Those numbers are listed in this column.

Camera Framing Camera framing directions for every proposed shot should make use of the symbols for basic camera shot framing: XLS, LS, MS, CU, and XCU. This will communicate to the camera operator the lens framing for the proposed shot. The framing choice for the shot is indicated on the master script. That framing direction is recorded here.

Character/Object The content of each framed shot is indicated here. The master script contains this information with every storyboard frame sketched (e.g., the character Henry might be indicated by the initial "H" under the master script storyboard frame. "Henry," then, should be entered in this column).

Camera Movement Camera movement directions indicate the kind of camera movement designed for achieving the proposed shot. Camera movement in a shot can be primary movement. (*Primary movement* is movement on the part of the talent in front of the camera.) This will require camera movement—pan, tilt, or pedestal—to follow the actor or actress. (This camera movement is called *secondary movement*—the movement of the camera itself. This camera movement also could be a dolly, truck, pedestal, or zoom.) The technical director should translate those directions from the director's master script into this column for the camera operator.

Special Instructions The director may indicate some special instructions in the master script or blocking plot for a particular shot or camera operator. Those instructions should be indicated here. The camera shot list is

camera specific, so an individual camera operator can be addressed in these special instructions.

• **Continuity Notes Form**

Production process Studio drama production

Responsibility Continuity person

Purpose To record during studio production the details of all elements of the production as an aid to reestablishing details for sequential takes. These notes serve to facilitate a continuity of production details so editing across takes is continuous.

Objective The continuity notes form is a record of all possible studio production details (e.g., set, actor/actress, properties, and dialogue) that must carry over from take to take and across edits in postproduction. The nature of production is such that, while videotaped takes are separate in production, the final editing of the takes must look continuous. The continuity notes form records specific details of all facets of production in order to reestablish the detail in subsequent videotaping sessions.

Notes on the use of the continuity notes form The form is designed to be used for each separate take in production. This means that for any single unit, with many takes, one continuity notes page should be used. In preparing for a lengthy production session involving many units, many copies of this form will have to be used. The continuity person should be someone who is very observant. The role demands attention to the slightest detail of the production that may have to be reestablished in another studio take. The role requires constant note taking.

Admittedly, the immediacy of replaying video on a studio playback monitor is a fast check on any detail in a previous videotaped take. However, while playback is fast, it is time consuming and is not a habit to which a director should become accustomed. Careful, detailed continuity notes are still a requirement in production.

Glossary

Continuity Person The name of the crew member responsible for taking continuity notes is entered here.

Teleplay Title/Author The title and author of the teleplay are recorded here.

Taping Date This notation specifies the production session during which the notes were taken.

Page of This entry indicates the expected number of continuity notes forms for the production and the page number of any single form.

Set A description of the set or the environment for the production session is required here. Note should be made of things that were moved (e.g., action properties) in order to be replaced.

Interior/Exterior These choices allow the notation of whether the set is considered indoors or outdoors.

Day/Night These choices allow the notation of time of day of the setting.

Ending Camera/ C1, C2, C3 This box quickly records the studio camera that made the final shot of the take being noted.

Ending Shot No. This box records the number of the last shot made. The shot numbers are found on the master script and on the camera shot lists.

Videotape No. The continuity person should learn from the videotape recorder operator what videotape stock is being used as the recording source tape and record the code number here.

Script Unit No. The number of the script unit being videotaped is recorded here.

Script Page No. For a script unit being produced with numbered pages, the page number of the script page being videotaped is recorded in this space.

Preceding Edit Unit No.: Cut/Dissolve/Wipe The number of the preceding unit to be edited to the current take is recorded here with the type of visual transition designed for it from the director's master script.

Succeeding Edit Unit No.: Cut/Dissolve/Wipe Similar to the previous entry, this records the sequential edit unit called for from the master script and the visual transition designed for the edit. This information will indicate the special requirements for continuity details. If a wipe or dissolve is called for, a second B-roll videotape will be required to save a generation of video and a change of videotape code will be required.

Actor(s)/Actress(es)/Costume/Make-up/Properties Notes This area should be used to verbally record all details noticed during production that may have to be reestablished for a subsequent take or unit of the script. Note should be made of any costume use (e.g., a tie off center or a pocket flap tucked in), make-up detail (e.g., smudged lipstick or position of a wound), or property use (e.g., a half-smoked cigarette in the right hand between the index and middle fingers or a fresh ice cream cone in the left hand).

Circle Takes (1 through 12) This area of the form facilitates recording each subsequent take of a unit. The best use of these boxes is to circle each number representing the take in progress until the take is satisfactorily videotaped.

End Slate The normal practice of indicating on the videotape leader what take of a unit is being recorded is to use the character generator slate. If a particular take is flubbed or messed up for a minor problem (e.g., an actor missing a line), the director might simply indicate that, instead of stopping taping and beginning from scratch, the crew keep going by starting over without stopping for the slate at the beginning of the retake. Since the slate was not recorded at the beginning of the retake, an end slate is used—the slate is recorded on videotape at the end of the take.

This note will alert the postproduction crew to look for a slate not at the beginning but at the end of that take.

Timer/Counter This area of the form should be used to record a consecutive stop watch time or the consecutive videotape recorder digital counter number reported by the videotape recorder operator. The time and counter number are a record of the length into the videotape where this take was recorded. Therefore, the stop watch or the counter should be set at zero when the tape is rewound to the beginning.

It is good practice for the videotape recorder operator to call the counter numbers over the intercom to the floor director to inform the continuity person at the beginning of every take or retake.

Reason for Use/Not Good One benefit of the continuity notes is to save the tedium of reviewing all takes after a production session to make a determination of the quality of each take. If detailed notes are made, a judgmental notation on the quality of each take is recorded and one postproduction chore is complete. Usually, a producer or director makes a judgment anyway in determining whether to retake a unit or to move on to a new unit. The continuity person should note in a word or two whether the take is good or not and the reason for that judgment.

Action Continuity details on any action during the take should be noted here. For example, the direction an actor or actress takes when turning or the hand used to open a door should be noted.

Dialogue An area of continuity interest can be the dialogue. Notation should be made of any quirk of dialogue used in a previous take that may have to be repeated in a subsequent take intended for a matched edit.

- **Videotape Log Form**

Production process Studio drama production

Responsibility Assistant director

Purpose To record from the control room the decisions across the takes and retakes of scheduled script units during production sessions.

Objective The videotape log form is designed to record and note quickly the sequential videotaped takes of each script unit during production. This log becomes a valuable resource to the producer and director in postproduction by indicating the order of takes on a source videotape as well as the designation of the quality of each take. This form should save the need for reviewing in detail every take recorded on the source videotape.

Glossary

Director This line lists the director of the teleplay.

Assistant Director This entry records the name of the assistant director, who is responsible for logging the videotape information.

Log Form No. The videotape log forms should be numbered consecutively for ease in reordering the forms before postproduction.

Teleplay Title/Author The teleplay title and author are entered here.

Taping Date This notation records the date of the production session.

Page of This notation indicates the expected number of log forms for the production and the number of each individual log form.

Script Page The assistant director records the page number of the script for the unit being recorded.

Script Unit This box records the script unit being videotaped.

Videotape No. This entry cross checks the videotape stock being recorded by the videotape recorder operator and assists in organizing resources for postproduction.

Set The name or label of the set being used for the current take is entered here.

Notes This space permits any additional notations relevant to the takes being videotaped during the production session.

Circle Takes (1 through 12) This area of the form records each subsequent take of a unit. The best use of these boxes is to circle each number representing the take in progress until the take is satisfactorily videotaped.

End Slate The normal practice of indicating on the videotape leader what take of a unit is being recorded is to use the character generator slate. If a particular take is flubbed or messed up for a minor problem (e.g., an actor missing a line), the director might simply indicate that, instead of stopping taping and beginning from scratch, the crew keep going by starting over without stopping for the slate at the beginning of the retake. Since the slate was not recorded at the beginning of the retake, an end slate is used—the slate is recorded on tape at the end of the take.

This note will alert the postproduction crew to look for a slate not at the beginning but at the end of that take.

Timer/Counter This area of the form is used to record a consecutive stop watch time or the consecutive videotape recorder digital counter number reported by the videotape recorder operator. The time and counter number are a record of the length into the videotape that this take was recorded. Therefore, the stop watch or the counter should be set at zero when the tape is rewound to the beginning.

It is good practice for the videotape recorder operator to call the counter numbers over the intercom to the assistant director at the beginning of every take or retake.

Reason for Use/Not Good One benefit of the videotape log is to save the tedium of reviewing all takes after a studio production session to make a determination of the quality of each take. If detailed log notes are made, a judgmental notation on the quality of each take is recorded and one postproduction chore is complete. Usually, a producer or director makes a judgment anyway in determining whether to retake a unit or to move on to a new unit. The assistant director should note in a word or two whether the take is good or not and the reason for that judgment.

• **Talent Release Form**

Production process Studio drama production

Responsibility Producer

Purpose To give the producer legal rights over the video and audio recording of an individual's performance.

Objective The talent release form is a legal document that, when filled out and signed by the talent, gives to the producer and producing organization the legal right to use both the video and audio recording of an individual for publication.

This form is especially necessary in the case of talent if the producing company may profit from the eventual sale of the video product. If talent contracts were completed and signed, talent releases were probably included. When talent volunteer their services or perform for only a nominal fee or other gratuity, a signed talent release is recommended. Generally any talent being featured in video and audio taping should sign a talent release form before the production is aired.

Glossary

Talent Name This entry should contain the name of the individual talent recorded on video and/or audio tape.

Teleplay Title This entry records the title of the teleplay.

Recording Location The location site or production facility where the video or audio recording is made should be entered here.

Producer The name of the supervising producer should be entered here.

Producing Organization The incorporated name of the producing organization should be entered here.

Note: The expression "For value received" may imply that some remuneration, even a token remuneration, be required for the form to be legally binding. When there is any doubt about the legal nature of the document, consult a lawyer.

• **Postproduction Cue Sheet Form**

Production process Studio drama postproduction

Responsibility Director

Purpose To prepare an editing cue sheet before postproduction editing. The entries on the postproduction cue sheet are drawn from the videotape log form of acceptable videotaped takes from the source tapes.

Objective The postproduction cue sheet is a stage of postproduction editing preparation in which the director (or producer or assistant director) culls from the videotape log form the acceptable videotaped takes and videotape number and lists these sources in chronological script unit order.

Glossary

Producer The name of the producer of the teleplay is entered here.

Director The director preparing the form is listed here.

Editor Should the director not do the editing, an assigned technical director/editor is listed here.

Teleplay Title/Author The title and author of the teleplay are entered here.

Date The date of the creation of the cue sheet is recorded here.

Page of Each postproduction cue sheet should be numbered consecutively and each page recorded with the respective page number as the portion of the whole, e.g., "Page 1 of 1," or "Page 2 of 5."

Script Unit This column records in chronological order (i.e., in the consecutive order of the master script) all of the script unit numbers.

Log Form Every log form was numbered consecutively. Coordinating each log form by number with its respective script unit provides ready access to other information needed during postproduction.

Tape No. The number of the videotape stock on which the script unit was recorded is entered here. This, too, will facilitate access to information.

Take No. This entry records the acceptable take number on the source videotape previously listed. This information was originally recorded on the videotape log form (and on the continuity notes form).

Time Code Notation is made from the SMPTE time code striped on the source tapes during or after videotaping for the take listed in the previous column. If time code is not used or is not known, the timer time or videotape recorder counter reading should be picked up from the videotape log form or continuity notes form. These readings should have been made from the beginning of the tape as an accurate indication of the position of the take from the front of the videotape.

Length If accurate timing of the length of individual videotaped takes were recorded, that time notation should be entered here. A sum of these entries should give the producer and director a ballpark idea of the overall length of the teleplay.

Comments This area of the form allows additional remarks to guide editing.

Check off Column The end column of squares is added as a convenience for checking off the units as they are edited onto the master videotape.

• **Titling/Credits Copy Form**

Production process Studio drama preproduction

Responsibility Producer, production assistant

Purpose To explicitly define and describe all of the character generator copy to be used for the teleplay production including slate, title, cast, and crew credits.

Objective The titling/credits copy form organizes and permits the design and control of all video screen text considered the title and beginning and ending credits for the teleplay. This form is very specific and detailed. It is an attempt to be accurate and thorough in the design and production of all character generator copy.

Glossary

Production Assistant While the producer is responsible for the content of the titling and credits copy, the production assistant can design and enter the copy into the character generator and record the text pages. The production assistant's name is entered here.

Date The date of the design of the titling and credits copy form is entered here.

Page of This notation indicates the expected number of pages of titling and credits copy and the number of each individual page. This preserves the order of credits over numerous pages of copy.

Slate This model aspect ratio frame labels the important data to be contained on the character generator slate to be used before every videotaped take during production. This information is the least amount of copy necessary. Some production operations may require additional copy. The production assistant should check with the producer or director on the data to be included for any production.

Record No. This entry permits a record of the assigned number for each data page entered into the character generator.

Suggested Video Screen Frames Each successive video frame is labeled with the text assigned to the screen. There is no standard of content or order in which the information should be presented on the screen. The frames suggested here account for the cast and crew defined in this text only. Other productions may use fewer frames and others more. The final determination of the number of frames belongs to the producer and director.

Order This entry permits noting of the order of presentation of screen text information. In these boxes, the sequential order of screen presentation should be noted.

Rolled/Speed These entries allow the choice to roll the screen text pages up the screen and to indicate the speed chosen at the character generator for the roll.

Advanced/Time These entries offer the choice to advance one screen text page at a time (i.e., a cut). The time indicates the time allowed for each screen page to be on-screen.

Font Style The production assistant designs or selects a particular font style, available on the character generator, for the screen text.

Font Size The choice of font size is also available on the character generator and must be decided upon. That choice is entered here.

Color Character generators also permit the colorizing of text fonts. That choice is indicated here.

Effects Font style effects can also be selected (e.g., border or drop shadow). Some of these effects may have to be generated at the switcher. The technical director will have to be alerted to those choices.

Record No. This entry records the page number assigned to the screen text entered for that page.

Front Time This notation records the time for this particular screen to be matted on the screen as measured from the front of the video production (i.e., timed from first video).

Dialogue Cue Some screen texts will be matted on the screen at a particular dialogue or music cue. That cue should be noted here.

Closing Credits These aspect ratio frames attempt to organize and suggest necessary or optional closing credits to be considered for the end of the teleplay. There is no standard way to present closing credits.

Note: Graphic boxes within these video screen frames are intended to indicate a suggested position for the corresponding text. They are only *suggested* positionings. If they are used, the text could be lettered in the boxes themselves. Otherwise, multiple generic frames are included for creative design and positioning by the production assistant. Because text accuracy is demanded (e.g., when personal names are being matted on the screen), it is highly recommended that the character generator text be *typed* onto the copy form.

• **Effects/Music Cue Sheet Form**

Production   Studio drama postproduction

Responsibility   Audio director

Purpose   To list in chronological order all sound effects and music to be mixed with the soundtrack in postproduction.

Objective   This form organizes by script units all sound effects and music required for the teleplay, listing their sources and time duration with the in-cue and out-cue for each.

Glossary

Audio Director   The name of the audio director responsible for the audio design of sound effects and music is entered here.

Director/Editor   The name of the director or technical director/editor is listed here.

Teleplay Title/Author   The title of the teleplay and its author are recorded here.

Date   This notation records the date of the preparation of this cue sheet.

Page   of   This entry logs the expected number of pages for this cue sheet and the number of each page of the form.

Script Unit   This column records the script unit number from the master script requiring some postproduction effects or music.

Effect   The specific sound effect required is listed here.

Music   The specific music required is listed here.

Source: Cut/Fade   This entry notes the recorded source from which an effect or music is to be recorded. The source may be a prerecorded cartridge, a cassette, or a record album. In addition to the source of the effect or music, notation must be made as to whether the source is to be cut in or faded in and cut out or faded out. Check marks should be made in the appropriate boxes in this column.

Length   This column records the duration of the required effect or music, measured from the succeeding in-cue to out-cue.

In-cue   This entry notes the dialogue or action cue at which the effect or music is to begin.

Out-cue   This entry notes the dialogue or action cue at which the effect or music is to end.

Glossary

Above-the-line Above-the-line refers to that division of television production personnel involving those who are nontechnical, e.g., the producer, the director, etc. (*See also* Below-the-line.)

Academy leader The academy leader is the first minute of video preceding the content video of a recorded videotape. It consists of 30 seconds of color bars and audio tone, followed by 20 seconds of the slate of the content of the program, then by 10 seconds of black screen, and, finally, by the opening of the videotaped program.

Action props Action properties are those moving objects used by actors and actresses. For example, an automobile or horse and buggy are considered action properties.

Ambience Ambience is any background sound (e.g., city traffic or an airplane flyover) in a recording environment.

A-roll An A-roll video is the primary videotape recording source. In a studio production videotape environment, the A-roll is the master videotape of the program. B-roll is the videotape source that is inserted into the A-roll videotape.

Aspect ratio frame An aspect ratio frame is the television screen proportional rectangle drawing, 3 units high by 4 units wide. Aspect ratio frames are drawings used in the design of storyboards.

Audience demographics Audience demographics is the sum of the individual traits of an audience (e.g., age, sex, education, income, race, and religion).

Audio perspective Audio perspective is the perception that longer video shots should have a more distant sound and closer video shots should have a closer sound. Audio perspective attempts to recreate the sound distance perception of real life.

Audio plot Preparing an audio plot is a preproduction task requirement of an audio director, which includes the judgment of type of microphone to record required sound, microphone holder for picking up required sound, and physical placement of a microphone for videotaping. (See the audio plot form.)

Audition An audition is the forum in which prospective talent try out for a part in a play or commercial. Auditions can be used for any on-camera talent role (e.g., anchor for a newscast, host for a talk show, or character in a commercial or teleplay). Prospective talent should attend an audition with a vita or resumé and a black and white glossy picture of him- or herself. During the audition, prospective talent may be asked to read a portion of a script, to improvise, or to characterize a situation or character.

Base light or baselight Base light is technically the minimum light level necessary to best operate video cameras. The amount of light differs according to type of tube(s) in the camera. Base light colloquially also refers to a basic light design of a studio set.

Below-the-line Below-the-line refers to that division of television production personnel involving those who are technical and technical facilities, e.g., camera operator, lighting director, the production switcher, etc. (*See also* Above-the-line.)

Bird's eye view A bird's eye view is the point of view of a set looking directly down on the set from above, noting the confines of the set and set properties. The view can also contain the cast and the cameras.

Bite A bite is a portion of a video or audio recording.

Blocking Blocking is that process by which a director physically moves participants (cast and cameras) to differing points within a location or set.

Blocking plot Preparing a blocking plot is the preproduction task of a director, which consists of making a bird's eye view drawing of a set or recording environment with major properties indicated. A director indicates with circles where the talent will be placed. The circles are combined with arrows to indicate movement of the talent. From a blocking plot, a lighting director can create a lighting plot and a camera operator can decide camera set-up and placement. (See the blocking plot form.)

Boom microphone A boom microphone is a microphone, usually directional, designed to be mounted and held above the person(s) speaking. A boom microphone must be aimed at the mouth of the speaker and raised and lowered depending on the framing of each camera shot. (*See also* Audio perspective.)

Breakdown A breakdown is a preproduction analysis of either a script or a storyboard. It is intended to separate scene elements from the script or storyboard and arrange them in the proposed videotaping order. A breakdown is a necessary component to the development of a production schedule. (See the script breakdown form.)

B-roll A B-roll is a second videotape source needed in production to perform some video effects involving two video sources during videotaping or editing, such as a dissolve or a wipe. The A-roll would be the primary videotape source into which a B-roll is inserted.

Cast call The cast call is that appointed rendezvous time at which the members of the cast—actor(s) and actress(es)—assemble before beginning production tasks.

Character generator A character generator is a video effects generator that electronically produces text on a video screen. The text that is recorded in the memory of the character generator is usually used for matting over a colored background or other video image.

Characterization Characterization is the process by which an actor or actress becomes the script character being played. Characterization includes not only costuming, make-up, and dialogue but also the mental and motivational state of the character.

Clearance Clearance is the process of securing the rights to use copyrighted material. Clearance is most often secured for the legal use of music.

Composition of a shot The composition of a camera shot indicates the subject and arrangement of a shot as framed in the viewfinder of a camera. It would indicate the person or object to be framed and the degree of the framing (e.g., CU or XLS).

Consumable props Consumable properties are those properties that are consumed as part of using them (e.g., a cigarette is smoked or food is eaten). Consumable props require constant replacing

Contingency Contingency is that percentage amount added to a subtotal of estimated costs in budget making. A common contingency amount is 15% (i.e., 15% of the subtotal of estimated budget costs is added to the subtotal itself as a hedge against actual costs).

Continuity Continuity is the flow of edited images and the content details of edited images from shot to shot. Continuity observation entails the close scrutiny of talent, properties, and environment during videotaping to ensure the accurate flow of edited images in post-production. (See the continuity notes form.)

Contrast ratio Contrast ratio is the proportion of light to dark areas in electronic video images or across lighted areas on a set.

Control track The control track is a flow of electronic impulses recorded on the edge of the videotape that serve as synchronization units for accurate videotape editing. They serve the same purpose as do the sprocket holes on film.

Copy Copy refers to any scripted text to be recorded on the audio track of a videotape or in the memory of a character generator for matting onto a video image or background.

Copyright Copyright is the legal right of an artist or author to the exclusive control of the artist's or author's original work. Copyrighted material is protected by law, and the public use of such materials must always be cleared by a producer from the owner of the copyright.

Credits Credits are those on-screen texts that list the names and roles played or performed by all members of the cast and crew of the production.

Crew call A crew call is the stated time for the rendezvous of the production crew members and is usually held at the videotaping site.

Cut-away A cut-away is a video of related but extraneous video content inserted into the primary video.

For example, video images of a hospital operating room (related but extraneous) would serve as a cut-away insert to a video of an interview (primary video material) with a doctor.

Cut-in A cut-in is a video of necessary and motivated video images to be edited into an established or master scene. For example, close-up shots (necessary and motivated) of two people in conversation serve as cut-ins to a long shot (master scene) of the two people walking and talking.

Cyclorama A cyclorama is the ceiling to floor material used as a simple backdrop for some television studio productions. Cycloramas may be made of a black velour to provide a solid black background or a scrim material that may be lighted with any colored light.

Decibel A decibel is a unit of sound that measures the loudness or softness of the sound.

Edited master An edited master is the final editing of a video piece from source tapes.

End board/End slate End board and end slate are terms for the slate when a videotaped take is slated at the end of a take instead of at the front. The end board is used when a take is redone without stopping the camera and recorder. End board use often occurs when an actor simply flubs a line.

External video signal The external video signal is the control room signal that carries the program or line of the video program. In some studio operations, that signal can be routed to the individual camera monitors and called up by camera operators on their cameras. Having access to the external video signal allows the camera operators to match the camera framing over their monitors to the camera framing as seen on the monitor of a camera on line.

Final audio Final audio is used to designate the last sound of copy or music in a video piece. It is a designated end point to measure the length of time of an edited master video piece. *(See also* First audio.)

Final edit The final edit is the completed master videotape of a project. It is usually referred to in contrast to the rough edit, which is a preliminary edit of the videotape.

First audio First audio is used to designate the first sound of copy or music in a video piece. It is a designated beginning point to measure the length of time of an edited video piece. *(See also* Final audio.)

Fishpole A fishpole is an extendible holder for a directional microphone. A fishpole, usually handheld, is extended into a set during dialogue for videotaping.

Foldback Foldback is the process of feeding the audio signal back into the studio for the convenience of studio personnel and on-camera talent.

Format A format is the listed order or outline of the content of a video product or program. Format is also used to indicate the genre of a television program (e.g., talk show or newscast).

Framing Framing is the composition and degree of image arrangement as seen through the viewfinder of a camera. The framing is usually described in terms of how close or far away the subject is perceived to be from the camera (e.g., a close-up versus a long shot).

Freelancer A freelancer is a person who works in the film or video field on a production basis as opposed to being a permanent, full-time employee of a company. Writers, producers, and directors are examples of freelancers.

Freeze A freeze is the appearance of holding the video image on the screen still. A freeze is also used for production cast members when they must remain without movement while continuity notes are being made.

Friction Friction, usually pan and tilt friction, refers to the amount of drag that the pan or tilt mechanisms produce in performing their functions. Adequate friction gives a camera operator control of panning and tilting motions.

F-stop The f-stop units, the calibrated units on the aperture of the lens of a camera, determine the amount of light entering the lens and falling on the pickup tubes.

Gel Gel (gelatin) are filters used in the control of light for videotaping. Gels are used on lighting instruments to filter light or change lighting temperature and on windows to change the temperature of light entering a videotaping environment.

Hand properties Hand properties are those objects needed (handled) by the talent. Examples of hand properties are a knife or a purse.

In-cue An in-cue is the beginning point of copy, music, or video screen at which timing or video or audio inserting is to occur. *(See also* Out-cue.)

Insurance coverage Insurance coverage is required for the use of certain people (e.g., underage children or movie stars), facilities (e.g., television studio or remote locations), and certain properties (e.g., horses or automobiles).

Intercom network This term stands for intercommunication network. An intercom network is that system of two-way communication through which television production crew members can interface with each other during production sessions.

Leader The leader is the beginning portion of audio or videotape that is used to record information about the subsequent video or audio. Most leaders (often called academy leaders) contain the record of the slate and an audio check (e.g., :30 of tone) with a portion of video black before the video content of the recording. Some leaders also contain a portion of the color bars.

Legal clearance Legal clearance is the process by which recording and reproduction rights are obtained to use copyrighted materials. Legal clearance must be obtained for copyrighted materials (e.g., music, photographs, or film) and for synchronization (putting pictures to copyrighted music).

Levels Levels are those calibrated input units of light, sound, and video that have to be set to record at

acceptable degrees of unity and definition for broadcast reproduction.

Lighting design The lighting design is the preproduction stage during which the mood, intensity, and degree of light for a videotape production are created. The lighting design is the responsibility of the lighting director. The lighting design can be done after location scouting is complete. (See the lighting plot form.)

Location A location is that environment outside a recording studio in which some videotaping is to be done, often referred to as the remote location or, simply, the remote.

Logging Logging is the term used to indicate the process of recording continuity details during videotaping.

Master script A master script is the copy of the script that contains the preproduction information for videotaping. The master script is usually the director's copy and contains the final version of the copy, the storyboard, and the blocking of the cast and action props.

Master tape A master tape is the videotape containing an edited video project.

Mike grip A mike (microphone) grip is the individual responsible for holding the microphone or a microphone holder during the recording of audio in a studio or on location.

Mixing Mixing audio tracks in videotape editing is the process of combining two recorded audio tracks into one audio channel. Mixing usually adds music or ambient sound to a track of voice recording.

Out-cue An out-cue is the end point of copy, music, or video at which point timing or video or audio inserting is to end. *(See also* In-cue.)

Pacing Pacing is the perception of timing of the audio or video piece. Pacing is not necessarily the actual speed of a production or production elements, but it is the coordinated flow and uniformity of sequencing of all production elements (e.g., music beat, copy rhythm, or video cutting).

Pad Pad (padding) is a term applied in television production whenever some flexible video, audio, black signal, or time may be needed.

Pickup pattern Pickup pattern designates the sound sensitive area around the head of a microphone. It is the area within which sounds will be heard by the microphone.

Pickup shots Pickup shots refers to the practice of videotaping additional studio set footage, beyond scripted units, to be used during postproduction editing. Pickup shots often will be used as cut-aways to cover difficult or mismatched edits discovered during editing. Pickup shot subject matter could be such shots as a close-up of hands, a set table, an unopened door, or an attentive actor.

Platform boom A platform boom is a large device for holding a microphone and extending it into a set during videotaping. Most platform booms are on wheels, need to be steered when moved, and require at least two operators—one to handle the microphone and the boom on the platform and the other to move the platform.

Plot   A plot is a creative or technical design usually used for blocking actor(s) and actress(es) within sets, designing sound coverage in audio production, designing the lighting pattern on sets, and listing all properties.

Postproduction cue sheet   A postproduction cue sheet is an editing cue sheet which lists successive videotape units selected from production source tapes to be edited onto the master videotape. All videotape units are listed in the order in which they were produced on the videotape log form. (See the postproduction cue sheet form and the videotape log form)

Preproduction script   A preproduction script is a copy of a script that is considered subject to change. A preproduction script should contain sufficient audio and video material on which to judge the substance of the final proposed project.

Production meeting   A production meeting is a gathering of all production personnel to review details of a production.

Production statement   A production statement is a simple, one sentence expression of the goal or objective of a production. The production statement is a constant reminder at all stages of production of exactly what is being accomplished and why.

Production value   A production value is any element or effect that is used when motivated to create an overall impact. Some examples of production values are music, lighting, and special video effects.

Program format   A program format refers to the content and order of that content in a television program. Format can also refer to the genre of the program (e.g., talk show or drama).

Property   Property is the term used for any movable article on a set. Properties can include such items as furniture, lamps, telephone, food, or bicycle. *(See also Hand properties and Action props.)*

Rate card   A rate card is the list of space, hardware, personnel, services, and the cost for production facilities.

Record button   The record button is a circular red insert plug found on the bottom of most videotape cassettes. Removing the record button serves as a safety check against recording over or erasing previously recorded video. The absence of the record button will allow playback but not recording.

Roll tape   "Roll tape" is the expression used to signal operation of the videotape recorder at the beginning of a take. A location director uses the expression as a sign of the director's intention to videotape a scene.

Rough edit   A rough edit is the result of a first postproduction editing session where accurate timing and tight edits are not required. A rough edit serves as a preliminary step to the final edit. A rough edit provides the opportunity to make final edit decisions on pacing and image transitions.

Royalty   Royalty is the share of proceeds from the publication of some work of art such as a book, a play, or music. The performance of such works requires financial payment to the owner of the work.

Scale model set   A scale model set is a miniature version, usually in a one foot is equal to one inch scale, of a set that is used to design blocking, lighting, and audio. Many decisions that would normally have to wait until the real set is constructed can be made based on the scale model.

Script unit   A script unit is that section of a script that designates a videotaping portion of a script and is equivalent to a scene in a larger act. It is any gratuitous unit that a director may define for production purposes. Many teleplay scripts number all script units consecutively in the right- and left-hand margins.

Secondary motion   Secondary motion refers to those movements that occur with the movement of the camera. Secondary movements include pan, tilt, dolly, truck, arc, zoom, pedestal, and boom.

Serendipity syndrome   The serendipity syndrome refers to those good and pleasant effects in television production that were unplanned and unexpected.

Set properties   Set properties are those properties which decorate a set. Set properties include furniture, knick-knacks, lamps, etc. *(See also* Hand properties, Action props.)

Shading   Shading in video production is the control of the iris of the lens of a camera. It is the process of setting and/or controlling the iris to permit light to hit or exclude light from hitting the camera tubes.

Shoot   Shoot is a slang term used to describe a videotape production session either on location or in the studio.

Shooting order   The shooting order is the order in which the script units will be shot and is indicated on the production schedule. Most often, the shooting order is determined by the availability of locations and actors/actresses.

Shooting units   Shooting units are those portions of a television script that are producible in one continuous videotape take. Shooting units in television are similar to short scenes in theater. In drama production, shooting units are determined by pages of a script or portions of a page.

Shot list   The shot list is a form created during preproduction on which a director indicates types and order of shots to be videotaped. A shot list differentiates between master or establishing shots and cut-ins, provides framing instructions, and indicates the duration of each shot by out-cue. Shot lists are camera specific with the shot list for every camera beginning with the count of one.

Slate   The slate is a video recording device that allows the labeling of the leader of each take. A slate usually records the title of the production, the producer and/or director, date, take number, and videotape code. The character generator, a blackboard, or a white showcard can serve as a slate. Slate also indicates the action of recording the slate on the videotape leader.

SMPTE time code   SMPTE (Society of Motion Picture and Television Engineers) time code is an electronic signal recorded on a secondary audio track of videotape to assist an editor in accurately creating a videotape edit. SMPTE time code records the hours, minutes, seconds, and frame numbers of elapsed time for each video frame.

Sound effects   Sound effects are those prerecorded sounds that simulate sounds in the real environment.

They are used to create a lifelike environment in the television studio and for background to a dialogue track in drama production.

Source tape The source tape is any videotape stock used to record video that will later be edited into a larger videotape project. Source tapes are edited onto a master tape.

Spike marks Spike marks are the result of spiking cast or properties during production. A common form of spike marks are colored adhesive dots placed at the blocked positions of actors and actresses. A different color is assigned to each cast member.

Spiking Spiking is the process of recording with some form of marking the blocked position of a cast member, a set property, or a studio camera.

Stand-up A stand-up designates the technique of a location news reporter when the reporter appears on camera standing in the foreground of the location. A stand-up may be used as a lead, a bridge, or a tag for a video piece.

Stock shots Stock shots are usually exterior scenes used as scene settings for studio drama production. Common stock shots include cityscapes, sunsets, and house exterior shots. Many stock shots are found in videotape libraries. Some may be shot specifically for a particular production.

Storyboard A storyboard is a series of aspect ratio frames on which are sketched the proposed composition and framing of each shot to be videotaped. Storyboard frames are numbered consecutively, and the audio copy associated with each proposed shot is recorded under the frame. Storyboards are considered essential to some video genres (e.g., commercials) and are encouraged as a quality preproduction stage for all genres.

Strike A strike is the final stage of a location shoot when all production equipment is disassembled and packed for removal and the shooting environment is restored to the arrangement and condition found upon the arrival of the production crew.

Striping Striping is the process of recording SMPTE time code or control track on videotape as a measure of videotape control during editing.

Sweetening Sweetening refers to the process by which audio and video signals are cleaned up and clarified electronically in postproduction. Sweetening audio means to filter out background noises such as hums and buzzes.

Synchronization rights Synchronization rights are those legal clearances in which a producer receives the right to use copyrighted music in a videotape production.

Take A take is a single videotape unit from the beginning to the end of recording. A take usually begins with a recording of the slate and ends with a director's call to cut. It is not uncommon to record many takes of an individual unit. Many takes may be required for one shot.

Talent Talent is the term used to designate any person who appears in front of a camera, including actors/actresses and extras. Even animals are referred to as talent.

Talent release A talent release is a signed legal document by which a producer obtains the right to use the image, voice, and talent of a person for publication.

Target audience Target audience is the designation of that subset of the public for whom a particular video piece is designed. Knowing a target audience permits a producer and director to make calculated choices of production values to attract and hold the interest of the targeted group.

Timing Timing is the process of recording the length of a video piece from first to final audio. First audio and final audio might be music and not a verbal cue. Some video pieces may have a visual cue at the beginning or end of the piece.

Titling Titling is the design and production of all of those on-screen visual elements that create the title of a video production.

Vectorscope A vectorscope is an oscilloscope used to set and align the color of images as they are recorded by the videotape recorder.

Videographer A videographer is a photographer working in video.

Wrap A wrap is the stage of a production when the director indicates that a good take has been videotaped and signals a move to another shot from the shot list. A wrap is distinguished from a strike.

Selected Bibliography

TELEVISION PRODUCTION TEXTS

Armer, A. *Directing Television and Film.* Belmont, CA: Wadsworth Publishing Co., 1986.

Blum, R. *Television Writing: From Concept to Contract.* New York, NY: Hastings House, 1980.

Blumenthal, H. J. *Television Producing & Directing.* New York: NY: Harper & Row, 1988.

Carlson, V., and Carlson, S. *Professional Lighting Handbook.* Stoneham, MA: Focal Press, 1985.

Fielding, K. *Introduction to Television Production.* New York, NY: Longman, 1990.

Fuller, B., Kanaba, S., and Kanaba, J. *Single Camera Video Production: Techniques, Equipment, and Resources for Producing Quality Video Programs.* Englewood Cliffs, NJ: Prentice Hall, 1982.

Garvey, D., and Rivers, W. *Broadcast Writing.* New York, NY: Longman, 1982.

Hubatka, M. C., Hull, F., and Sanders, R. W. *Sweetening for Film, and TV.* Blue Ridge Summit, PA: TAB Books, 1985.

Huber, D. M. *Audio Production Techniques for Video.* Indianapolis, IN: Howard Sams & Co., 1987.

Kehoe, V. *Technique of the Professional Make-up Artist.* Stoneham, MA: Focal Press, 1985.

Kennedy, T. *Directing the Video Production.* White Plains, NY: Knowledge Industry Publications, Inc., 1988.

Mathias, H., and Patterson, R. *Achieving Photographic Control over the Video Image.* Belmont, CA: Wadsworth Publishing Co., 1985.

McQuillin, L. *The Video Production Guide.* Sante Fe, NM: Video Info, 1983.

Miller, P. *Script Supervising and Film Continuity,* Second Edition. Stoneham, MA: Focal Press, 1990.

Millerson, G. *Video Production Handbook.* Stoneham, MA: Focal Press, 1987.

Nisbett, A. *The Use of Microphones,* Second Edition. Stoneham, MA: Focal Press, 1983.

Schihl, R. J. *Single Camera Video: From Concept to Edited Master.* Stoneham, MA: Focal Press, 1989.

Souter, G. A. *Lighting Techniques for Video Production: The Art of Casting Shadows.* White Plains, NY: Knowledge Industry Productions, Inc., 1987.

Utz, P. *Today's Video: Equipment, Set Up and Production.* Englewood Cliffs, NJ: Prentice Hall, 1987.

Verna, T., and Bode, W. *Live TV: An Inside Look at Directing and Producing.* Stoneham, MA: Focal Press, 1987.

Weise, M. *Film and Video Budgets.* Stoneham, MA: Focal Press, 1980.

Wiegand, I. *Professional Video Production.* White Plains, NY: Knowledge Industry Publications, Inc., 1985.

Zettl, H. *Television Production Handbook.* Belmont, CA: Wadsworth Publishing Co., 1984.

Zettl, H. *Sight, Sound, Motion: Applied Media Aesthetics.* Belmont, CA: Wadsworth Publishing Co., 1990.

DRAMA PRODUCTION

Chamness, D. *The Hollywood Guide to Film Budgeting and Script Breakdown.* Toluca Lake, CA: D. Chamness & Assoc., 1977.

Dmytryk, E. *On Screen Directing.* Stoneham, MA: Focal Press, 1984.

St. John Marner, T. *Directing Motion Pictures.* New York: A. S. Barnes, 1972.

Trapnell, C. *Teleplay.* New York: Hawthorn Books, 1974.

Index